T0205493

Practical Guidelines for the Chemical Industry

Kiran R. Golwalkar • Rashmi Kumar

Practical Guidelines for the Chemical Industry

Operation, Processes, and Sustainability in Modern Facilities

 Springer

Kiran R. Golwalkar
Consulting Chemical Engineer
Nagpur, India

Rashmi Kumar
Assistant Professor of Chemical
Engineering
Dwarkadas J. Sanghvi College
of Engineering
Mumbai, India

ISBN 978-3-030-96583-9   ISBN 978-3-030-96581-5   (eBook)
https://doi.org/10.1007/978-3-030-96581-5

This Springer imprint is published by the registered company Springer Nature Switzerland AG
The registered company address is: Gewerbestrasse 11, 6330 Cham, Switzerland

This work is dedicated to the designers of chemical plants; as well as technicians and engineers working in chemical industries and the managements who will use the guidelines for their safe, pollution free and efficient running; and the teachers who will use them to train their students for successful careers in these industries.

Preface

Safe, pollution free and efficient running of chemical industries is vital for the economy of a nation and getting employment for the population. Chemical industries produce refinery products and petrochemicals (such as diesel, petrol, lubricants, ethylene, propylene, benzene); inorganic chemicals (sulphuric acid, oleum, chlorine, caustic soda, fertilisers, alum, paints, titanium dioxide, etc.); products such as polyethylene, polypropylene, polyvinyl chloride; etc. They also produce large quantities of specialty chemicals such as catalysts, surfactants, specialty polymers, coating additives, and oilfield chemicals which are used for specialised applications; synthetic fibres for clothing; paper for printing and writing, lubricants for machineries, agricultural chemicals (such as herbicides, fungicides, and insecticides); food preservatives; and active pharmaceutical ingredients which are necessary for ensuring food and health care for the nation. A large world market is generally available due to chemical industries.

The managements of these industries will have to consider production of quality products, expansion of capacities, diversification to new products, improvement of operational efficiency, modernization and revamping in certain cases, and acquisition of idle plants in order to maintain and improve operations, attain higher market share, and build up reputation in international markets.

Chemical industries have many process units and machineries which are to be operated (mostly continuously) in a carefully controlled manner. The operating conditions and process parameters are to be controlled by proper feed rates of raw materials, heating/cooling of the reactors, flow of absorbing liquids and gases to the absorption towers, and operating the process units in a safe, energy efficient manner without causing environmental pollution.

Incorrect control of operations can result in loss of production and unsafe conditions (*may even result in accidents*); environmental pollution, poor quality of products, higher consumption of raw materials and energy; and higher cost of maintenance due to more breakdowns of process units and machinery. Apart from these, the products may even get rejected in the market and the manufacturing organisation can run into a loss. These can have a negative impact on the organisation.

This book provides practical guidelines to correctly select a safe and efficient process for manufacture, as well as process units which need to be designed, fabricated, operated, and maintained properly to work efficiently without causing environmental pollution in order to make available quality products at reasonable costs to the customers.

It also gives general practical guidelines for important matters like operational safety, use of pressure vessels, selection of appropriate materials of construction for process units, efficient handling of process fluids, using proper heating and cooling systems, disposal of dangerous wastes, etc. These guidelines are therefore applicable to most of the chemical industries for sustainability in modern facilities.

Engineers and technicians working in industries like petroleum refining, fertilisers, chlor-alkali manufacture, synthetic fibres, etc. can use the basic information provided in this book along with manuals and instructions for the specific industries.

Nagpur, India   Kiran R. Golwalkar
Mumbai, India   Rashmi Kumar

Who Can Benefit from This Book?

This book has been written and the information in it is compiled for those working in chemical industries and for those interested in academics. It will be found useful for successful careers in units manufacturing various chemicals, fertilisers, and pharmaceuticals, for pollution control, etc.

Guidelines are given for establishing production units, their safe and smooth operations, for maintaining, expansion, and modernization, and related important activities such as:

- Statutory compliances
- Selection of process technology and equipment
- HAZID, HAZOP, and ensuring safety
- Management of human resources
- Pressure vessels
- Selection of process piping and pumps
- Selection of Materials of Construction for various units
- Selection of heating and cooling systems and equipment
- Cogeneration of steam and power
- Reducing breakdowns of equipment
- Reducing cost of production
- Process control
- Project activities for expansion, modernization of existing plants, revival of idle plants, relocating a plant
- Facilities for Effluent Treatment and Air Pollution Control
- Disposal of corrosive, toxic, inflammable waste

It is therefore intended to serve as a reference book for chemical and other plant engineers as well as other technical personnel, for managements of chemical industries as well as students of chemical and mechanical engineering. It is hoped that the book will be found useful by all of them. It has been written and compiled after getting inputs from academic personnel and industrial professionals.

Suggestions from readers for improvement of the book are welcome.

Sources of Our Information

This work is based on the knowledge and experience gained during working in chemical industries in India and abroad for manufacture of sulphuric acid, carbon-di-sulphide, oleums, alums, organic dyes, and destruction of hazardous wastes. The plants for these were installed in GRASIM Industries (India), Gopal Anand Rasayan (India), V K Engineers (India), EAHCL (Kenya), Thai Rayon (Thailand), Parksons Dyestuff Industries, and SMS Infrastructure (India).

Experience was also gained during erection, commissioning, and operation of these chemical plants, their modernization, and troubleshooting by the authors. These chemical industries had various process units such as refractory lined furnaces, hot gas filters, liquid sulphur filters, catalytic converters, waste heat recovery boilers and economisers, absorption towers, refractory lined electric furnaces, distillation columns, refrigeration units, effluent treatment plants, air pollution control facilities, heat exchangers for various duties, etc.

There used to be detailed discussions and regular interactions with the personnel at these industries on matters related to procurement of equipment, erection, safe operation, environmental pollution control, capacity expansion, and modernization. The knowledge and information received from senior managements, operating and maintenance engineers, and other personnel in these industries was very useful for the present work and is gratefully acknowledged.

The authors were also engaged in carrying out teaching assignments at universities for graduate level students for subjects such as Inorganic Chemical Technology, Process Equipment Design, Mass Transfer Operations, Plant Erection and Commissioning, and Environmental Pollution Control which are related to chemical engineering.

Various standard textbooks, *Perry's Chemical Engineers' Handbook*, *Coulson and Richardson's Chemical Engineering Design*, Vol 6, R. K. Sinnott, web sites of equipment manufacturers and Engineering Toolbox, our own study notes, and books authored by the same author and published by Springer International were referred while writing and compiling information for this book.

Acknowledgements

We are grateful to **Mr. Michael Luby,** Senior Editor Physical Sciences & Engineering, **Springer Nature**, for the guidance and support, and permission to refer to books published by Springer International earlier and use some relevant matter from them for the present book.

We are also grateful to the expert reviewers for the valuable time spent by them on giving the constructive feedback for improvement of the book. We also thank Ms. Sanjana Sundaram, Project Coordinator (book production), Springer Nature, and her team for the guidance and support provided for completing the manuscript.

We sincerely thank Dr. Sonali Dhokpande for her contribution to the chapter on Pressure Vessels; we also thank **Nandu Printers and Publications (Mumbai)** for permitting to use in this chapter the figure of Heads of Pressure Vessels from the book *Process Equipment Design and Drawing* authored by Mr. Kiran Hari Ghadyalji.

We sincerely thank **V K Engineers**, MIDC Tarapur, India (manufacturer of equipment for chemical plants), India office of **TLV International** (manufacturers of steam traps and related equipment) for permitting to use information for heat exchangers, jacketed reactors, and steam traps, and Editor of *Engineering Toolbox* for permitting us to use information published by them for properties of fuels, materials of construction, etc.

We would also like to thank Mr. Sagar Patel, Mr. Ganeshraj Baikerikar, Mr. Sanjay Deshmukh, and Mr. Chirag Thakur for their assistance in compiling data, useful suggestions, and making drawings for the present work.

We also thank Ms. Tessy Priya, Project Manager for her sincere efforts to carry out the corrections in the book proof as required by us and for the support provided.

Kiran R. Golwalkar and Rashmi Kumar

Contents

About the Authors

Kiran R. Golwalkar is a consulting chemical engineer in private practice in India. He worked for many years in the field of commissioning, operating, modernising, and diversifying of chemical plants in India, Kenya, Thailand, and Indonesia. He has also served as a visiting faculty member in chemical engineering institutes in Mumbai, Nagpur, and Pune in India.

Rashmi Kumar is Assistant Professor of Chemical Engineering (D J Sanghvi College of Engineering, Mumbai) and has a long teaching experience of about 25 years.

Chapter 1
Management Functions

Abstract Chemical industries are very important for the economic progress of a nation. They are also very useful for a comfortable life of the citizens, since they assist in meeting the requirements of food, clothing, and health products. Modern Management and responsible engineers in chemical industries shall understand and execute important activities necessary to establish, operate, and maintain these industries in a safe, pollution-free, and an economically efficient manner. Research and development activities shall be carried out for the progress of the organisation.

Keywords Successful managements · Arranging funds for various activities · Market survey · Selection of process and technology · Procurement of plant units and machinery · Manpower planning · Water budgeting · Efficient maintenance · Aims for research and development

1.1 Introduction

Chemical industries are very important for the economic progress of a nation and for meeting the requirements of food, clothing, and health products for ensuring a comfortable life for the citizens. Modern Management and responsible engineers in chemical industries shall always aim to operate and maintain these industries in a safe, pollution-free, and efficient manner, while manufacturing the products as per specifications at a reasonable cost.

The revenues obtained should be judiciously allocated for ensuring safe operations, environmental pollution control, energy efficiency of the industry, and for procurement of all necessary inputs like raw materials, spares for machineries for steady production, and packaging items for safe and proper transport of finished products to the customers.

Reasonable allocation shall also be done for innovation in working, improvement in existing technology, and development of new products and processes.

1.2   Successful Managements

The Promoters and Board of Directors shall proceed for Formation of Management team with experts from different fields. These experts shall be professionally qualified, experienced, and trustworthy individuals for carrying out various activities required for establishing the industry. They should be familiar with the typical nature of chemical plants which are generally run continuously and are prone to accidents, (there are exceptions also); they have chances of causing environmental pollution and are dependent on steady supply of water and power. Managements have to address the following issues for successfully establishing and running the chemical industries. This book gives guidelines for looking into these matters (given as check points in the next few chapters).

- Arranging funds for various activities
- Market survey (to understand requirements of quality and quantity by the customers).
- Selection of process and technology for meeting the above requirements.
- Getting statutory permissions and consent to establish and to operate (please see Sect. 9.1 for details)
- Procuring plant units and machinery.
- Planning for skilled and unskilled workforce.
- Arranging site infrastructure; supplies of water, power, raw materials, fuels, and fire fighting equipment.
- Erection and commissioning of process plant; and stabilising the production runs.
- Efficiently and safely operating the chemical plant without causing environmental pollution while ensuring manufacture of the products as per specifications and delivery schedules given by customers.
- Maintenance of process plants and auxiliaries to ensure safe and smooth working.
- Generating sufficient revenues to meet expenses and getting reasonable profit.

Additionally, as a part of Corporate Environmental Responsibility (CER) and Corporate Social Responsibility (CSR) the organisation must adopt practices and procedures to control pollution, eliminate waste and emissions, conserve forests, have funds allocated for rehabilitation and resettlement of displaced persons, sanitation, health, education, and skill development. All CER activities should aim to mitigate any damage to environment caused by the company and at the same time maximise the efficient use of resources.

1.3   Arranging Funds for Various Activities (While Minimising Capital Requirement)

Investment in plant and machineries as well as working capital are required for various activities such as market survey, land acquirement, obtaining licenses for various clearances; designing, fabrication, and procurement of machinery; construction of the site infrastructure, erection and commissioning of plant units, arranging supply of raw materials and utilities; effluent treatment, air pollution control, and disposal of dangerous wastes; product marketing (storage, quality check, and transportation) and employing manpower.

Requirement of funds for these activities are to be estimated and arrangements are to be made for the same. This can be done through own capital investment, floating equity shares, preference shares, and debentures; getting loans from banks and financial institutions; and bonds with fixed interest, which are some of the means for raising funds. Management can also explore the possibilities of getting government grants or subsidies (for setting up the industries in certain areas or for certain products for export) and raising funds through venture capital, which can reduce the amounts of loan to be taken. Requirement of capital can also be reduced if some second-hand (old) equipment are available in usable condition, which can be procured at lesser cost. They may be repaired or modified to suit the requirement.

Financial consultants may also be engaged for further advice in these matters.

Management shall carefully monitor all activities in order to establish and run a well-equipped plant to ensure production of a better product at minimum cost.

1.4   Market Survey

Market survey is to be carried out for a correct assessment of consumers' preferences; and an investigation into the state of the market (*for the product being considered for manufacture*) is to be done to find out the availability of similar or alternative products (their quality and unit costs) from competitors. The survey should obtain correct information about any special needs of the customers for quality, *type of packaging generally preferred* by the customers, and the consumption patterns. It can be done by collection of samples from other suppliers or buying of small quantities for analysis in own laboratory to confirm their quality. The tolerance limits for impurities (as mentioned in the specifications given by customers) shall also be carefully looked into.

It helps to identify the requirements and opportunities in the market. This plays a key role in estimating the demand for the product with required specifications. It helps to decide the manufacturing capacity of the plant for the product (with the particular qualities that have to be ensured during production). The plant is to be run accordingly to meet the specified quality and demand in the market.

Sale of Products
Cost of manufacturing the product will depend on the special characteristics/quality of the product, production capacity planned and quantity as well as delivery schedules (as specified by clients), efficiency of the process for manufacture, and costs of various inputs.

Cost of the delivered product (landed cost at delivery point) includes cost of special containers, safe transportation, *cost of unloading at points of use,* and transit insurance. Safe handling and dispatch of products must be ensured by proper packaging and careful transport to points of use at a reasonable (competitive) cost. These costs are generally borne by the supplier.

However, the selling price depends on the terms and conditions of sale and a reasonable profit margin required by the management.

1.5   Process Selection

It is advisable to choose from a variety of processes (which are feasible and available) for the manufacture of a product. Some of these processes could be based on drastic operating conditions of pressure, temperature, handling of corrosive or dangerous materials, high concentrations, etc. These shall be considered for initial evaluation of processes for shortlisting. It should be an industrially proven process as far as possible, except when a new process developed in the laboratory is to be adapted.

Example   *65% Oleum can be manufactured by adding 25 % oleum to liquid SO_3 in a circulation tank **or** by absorbing SO_3 vapours in a circulating stream of cooled 65% oleum to which 25% oleum is added at a controlled rate. The concentration is maintained at 65% and the product is transferred to storage. The latter method is safer as handling of liquid SO_3 (which is dangerous) is not required.*

The process to be used for manufacture of the products should be safe to operate, should have less drastic conditions of operations of pressures and temperatures (i.e. not necessary to use very high temperatures or pressures), shall not require handling of dangerous (toxic / inflammable) materials as far as possible, shall not have many side reactions/by products (which can reduce the yield of the main product), and generate minimum effluents.

HAZID study shall be carried out before final selection of the process
It shall normally result in a high degree of conversion of raw materials into the product (with minimum need for further concentration or purification). The technical team in the organisation must be familiar with the process.

Managements shall carefully study the Basic Engineering Package (BEP) prepared by their own technical team or by the consultants appointed. The BEP should give process description, process flow sheet, rated production capacity, broad specifications of equipment, typical plant layout, time required for implementation, manpower required, approximate consumptions of raw materials and utilities, etc. to enable estimation of economic viability. This shall be confirmed by estimation of competition i.e. overall production statistics and product specifications of competitors.

1.6   Selection of a Proven Technology

After a proper study of the BEP and its acceptance by the technical team, the senior management shall instruct preparation of Detailed Engineering Package (DEP) . This should be based on a technology which should be a proven one with plants of comparable capacity operating successfully for at least three years. These shall be based on raw materials which should be regularly available at reasonable costs. If the raw materials are not available as per specifications, additional equipment shall be included in the plant technology for their purification (by drying/calcinations/ filtration, etc.).

The DEP must give detailed mass and energy balance, detailed fabrication drawings and detailed specifications of all equipments (including operating conditions and materials of construction), P & ID, Proposed layout for the plant, enquiry document sheets for various equipments, procedures for erection and commissioning, operating and maintenance procedures etc.

The plant design should have scope for future expansion of capacity, diversification to more products, and should be energy efficient.

The plant must be safe to operate and maintain; should be suitable for operation at a capacity lower or higher than rated capacity i.e. should have sufficient turn down ratio while meeting product quality as per client requirements; and without causing environmental pollution when the plant is run at a little overload (say, about 20% above the rated capacity) in order to meet sudden high demand in the market from a very important customer or for export. The plant design should have flexibility of operation (say, about 20% below the rated capacity of production also) even when certain raw materials may not be available exactly as per required specifications, or are in short supply.

The conversion of (yield) raw materials shall be maximum, thus reducing the requirement of and expenses on raw material. Requirements of utilities and other consumable inputs (e.g. property modifiers, stabilisers, and spares for machineries etc.) shall also be minimum.

The technology used should not cause environmental pollution (generation of effluents should be minimum and those which can be easily taken care of). This minimises expenses on effluent processing. It will be easier to comply with the norms prescribed by the statutory authorities for pollution control when such technology is used.

Plant equipment should be able to withstand adverse weather conditions.

Handling of raw materials, operation of process units and machinery, and dispatch of finished products shall be mostly automatic. This can reduce requirement of labour for the plant.

- Such technologies can minimise/reduce the cost of production and hence the product can become competitive in the market.
- There shall be no infringement of patents. The technology should not violate any statutory rules and regulations.

1.7   Procurement of Plant Units, Machinery, and Their Installation

The proposed layout of the process plant as given in the DEP should comply with the Statutory Rules and Regulations for safe working of personnel in the premises as well as surroundings. It should be designed to safely store all dangerous, corrosive, inflammable, and toxic materials that may have to be handled.

After acceptance of BEP and DEP inquiries are to be floated for purchase/supply of equipment required for the plant. Offers are to be obtained for each item of equipment from at least three parties. Vendors shall then be shortlisted on the basis of their technical capability, past experience, and timely delivery. Final purchase orders are issued after detailed technical and commercial negotiations. The technical team may allow certain deviations to the vendor/fabricator if they make good suggestions for improvement in the design. The final design can include such small changes. Genuine difficulties in fabrication can also be considered.

Certain costly equipment or machinery could also be taken on lease if they can considerably increase the efficiency, but funds are not available at present.

Example A waste heat recovery boiler (instead of an air-cooled heat exchanger for process gases) can make steam available without consumption of fuel, but will be quite costly. It can be taken on lease from a party and the cost can be paid through savings generated by using the steam that can become available as a by-product.

HAZID and HAZOP studies should be carried out while selecting the process, technology, and during procurement, installation, and operation of these equipment to ensure safe working of the plant *without causing environmental pollution.*

1.8   Manpower Planning

The organisation structure shall clearly define the relationship between senior executives, plant manager, shop floor engineers, operating and maintenance personnel as well as other departments like marketing, stores, administration, research and development, materials procurement, and human resources. An organogram shall be created for the smooth functioning of the organisation.

Chemical industries need skilled manpower (engineers to build and manage the plant), operators and technicians (to operate and maintain the process units and machinery); scientists (for research and development purposes), experts for financial planning, cost accountants (in order to study various costs incurred), and marketing personnel (for market survey and final product marketing).

The persons to be engaged for skilled jobs need proper understanding of the process and machinery, their operating methods and maintenance procedures.

Sufficient manpower shall be engaged if the operations are complicated or involve many steps; if there are considerable physical distances between process units, machinery, and control systems; and as leave relievers.

Semiskilled workers should be very cautious and remain alert always. Their work includes unloading of bulk raw materials and chemicals required for manufacture; packaging finished products and loading for safe transportation.

Unskilled employees are also required for the performance of simple duties, which could be activities of a routine and repetitive nature although familiarity with the occupational environment is necessary. Their work may thus require, in addition to physical exertion, familiarity with variety of articles or goods. They are usually employed under contract for transportation, loading, and unloading under supervision, and cleaning or assistance for maintenance.

Careful planning of safe work environment, safety devices, and equipment is necessary for the process plant with proper protection and appropriate working and rest hours for workers.

Managements shall also consider carrying out skill upgradation programmes for the workers through training courses. Refresher courses can also be arranged for the personnel. Such activities can increase the efficiency of the work force. They can improve the safety, pollution control, product quality, and organisational efficiency.

1.9 Arranging Site Infrastructure

Management should arrange site office, security, firefighting equipment, communications with head office; and proper facilities for the storage of raw materials, all necessary spares; and make arrangements for sufficient power, water, and fuels for setting up the plant and for the smooth operation.

1.10 Erection and Commissioning

This can be carried out after getting statutory permissions for establishing the plant, followed by procurement of process units and plant machinery, making arrangement for site infrastructure, and engaging appropriate manpower/external contractors

1.11 Utilities Planning

Utilities and energy are major contributors to expenses for production by the plant. The cost of utilities can be reduced by proper plant design to minimise power, steam, and water consumption, by maximising efficient energy recovery, and optimising co-generation of power and heat, optimum use of compressed air, and careful use of auxiliary equipment which consume power.

1.12   Water Budgeting

Water is required in a process plant for the formation of products (by chemical combination), for the crystallisation of the products, for making solutions, for generation of steam, make up to cooling towers, flushing of process units, etc. The requirement of fresh supply of water for the plant should be minimised by reducing the consumption wherever possible and minimising wastage. This can be done by recycling of condensate to boiler feed, treatment of waste water and maximising its recycling for process, scrubbing systems, as make up to cooling towers, etc. Arrangements for rain water harvesting should also be installed. These actions can reduce the need for fresh supply from external sources as well as the cost of treatment of raw water.

Treated wastewater that cannot be reused for process or as make up to cooling towers shall be used elsewhere in the premises. Zero Liquid Discharge is now mandatory, and all efforts should be done for ensuring this. If this cannot be done even after best efforts, the effluent shall be disposed through Common Effluent Treatment Plants (CETP) statutorily authorised. However, conditions laid down by the CETP are to be complied with before the effluent is accepted by them for treatment (Fig. 1.1).

1.13   During Regular Operations of the Process Plant

Following matters are to be monitored regularly by plant engineers (both operating and maintenance engineers) and senior management for the safe, pollution-free, and efficient working of the plant. The marketing department can also provide useful inputs for product quality and manufacture.

Essential check points are given when the plant is operating (guidelines for plant engineers), and for maintenance (guidelines for maintenance engineers) for the following:

- Compliance with all statutory matters and instructions from government and local authorities e.g. for pressure vessels, storages of fuel and dangerous items; electrical installations and maximum demand on power supply grid to name a few.
- Proper operation and maintenance of effluent treatment plant (ETP) and air pollution control (APC) units; safe disposal of solid/hazardous wastes; and ensuring sufficient stocks of neutralising and treatment chemicals to meet instructions from pollution control authorities
- Meeting specifications and delivery schedules for the products

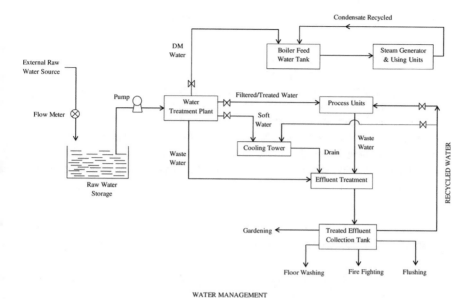

Fig. 1.1 Water management for a chemical plant

- Proper production planning in the case of batch production plants.
- Thorough cleaning and maintenance of process units and important machinery (pumps, blowers, reactors, condensers, and heating and cooling systems.)
- Minimising cost of production by controlling consumption of all inputs such as raw materials, catalyst, property modifiers, utilities and fuels, packaging items, and manpower.
- Inventory control of all essential inputs and spare parts.

1.14 Maintenance of Process Units and Machinery

- Proper maintenance is essential to make the plant safe, to reduce pollution, to increase yield from inputs, and to improve product quality. It is advisable to establish necessary facilities for maintenance in the beginning itself, since there could be minor requirements for repairs during commissioning and while stabilising the plant operations.
- Efficient management of maintenance constitutes the following: (1) preventive activities to minimise breakdowns, (2) timely repairs and/or replacement of damaged units, (3) thorough inspection, cleaning, servicing, and internal repairs of important equipment during annual shutdown for overhaul, (4) inventory control of spares and proper storage of all other inputs, (5) developing spares in-house, (6) reconditioning of certain machinery to reduce dependence on external agencies, and (7) better training for operating and maintenance personnel.

1.15   Aims for Research and Development

- Improvement in product quality
- Innovation in the process plant itself for making it safer; to minimise environmental pollution and to make it more energy-efficient.
- Reduction of consumption of raw materials and utilities.
- Reduction of maintenance problems.
- Development of new products, better effluent treatment, efficient energy recovery, use of cheaper raw materials, better design and materials of constructions (MOC) for critical parts, regeneration of spent catalyst, and other consumables if possible.

References/for Further Reading

1. "Process Equipment Procurement in the Chemical and Related Industries" Kiran Golwalkar, Published by Springer International Publishing, Switzerland.
2. "Production Management of Chemical Industries" Kiran Golwalkar Published by Springer International Publishing, Switzerland.

Chapter 2
Hazid, Hazop, and Ensuring Safety

Abstract It is essential to recognise the hazards and chances of accidents in a chemical industry before establishing and operating it. A detailed study shall be carried out of all materials that will be handled, the reactions and operations, and the various equipment that will be installed in the plant. Technical teams shall be formed for such detailed study. It should have knowledgeable and experienced designers, operating and maintenance personnel from different disciplines. They shall give suggestions after study of the P & ID for safely running the plant operations.

Keywords HAZID · Hazard analysis · Classification of hazards · Hazards identification · Prevention of dangerous situations · Ensuring safety in plant · HAZOP (hazard and operability) study · Formation of a technical team · Case studies

2.1 Introduction

For a proper and rational plant design it is very important to recognise the dangers involved in a chemical project before making any substantial investment in procurement of process units and machinery. It is necessary to consider the safety issues also. The technical team appointed by the management shall recommend appropriate preventive measures and safety factors during process and equipment designs for minimising hazards such as fire, explosion, and personal hazards from inflammable and toxic materials.

2.2 Hazard Analysis: HAZAN

This shall be based on a thorough study of the production process conditions, the reactions that will take place, properties of materials (and a study of their Material Safety Data Sheets) that will be handled, the generation of dangerous effluents, the quantities stored at site, and an examination of the probability of occurrence of dangerous incidents

Details of all main equipment in the plant, piping, control systems and the auxiliary facilities provided should also be studied to visualise the chances of occurrence of hazardous events.

It is imperative that the equipment are designed, fabricated, and tested as per applicable codes and standards and as per good engineering practices. The plant layout shall be designed for safe working. Appropriate instrumentation, control systems, and safety devices shall be provided as per statutory rules and checked regularly.

Factors such as improper plant design, faulty organisation of jobs, employing insufficiently skilled workers for a job (which needs skilled persons), unsafe working conditions, poor lighting, and dusty and noisy environment can also make the workplace hazardous. Hence the management should prioritise operational safety and health of workers; and improve the efficiency of both the workers and the plant.

Detailed Study

The detailed study will enable identification and analysis of the more likely causes and consequences of the hazardous incident that may occur.

Thereafter estimate the consequences to employees (persons present in the premises); people who are present nearby but outside the premises and the surrounding environment away from the place of incident.

A visit may be undertaken to a similar operating plant to observe the safety precautions and important features of plant equipment designs.

2.3   Check Points for HAZAN Study

- Analyse the hazards in the process, equipment, and operation of the plant.
- Identify the likely causes/mistakes during operation/unintended outcomes of operation that may result in development of dangerous situation.
- Identify ways to minimise process and equipment hazards due to corrosion, fire, explosion, and personal hazards from fumes and poisonous materials.
- Prevention and control: This refers to that phase of process design which can minimise/control the fire until available extinguishing facilities can work effectively for complete extinguishment of the fire.
- Assess the risks from such dangerous situations and whether the consequences will be severe (loss of lives) or quite damaging (loss of production due to major breakdowns).
- Suggest, evaluate, and plan the detection, prevention, and damage control measures.

2.4 Classification of Hazards

The following types of hazards can occur in a chemical plant.

- Fire hazards: This is the main hazard in chemical plants that can result in personal loss, property damage, and production interruption. Fire hazards can be minimised by preventing open flames, proximity of inflammable vapours to hot surfaces, electrical short circuits/sparking, and leaks of inflammable materials from any source. Fire protection can be achieved by regular maintenance of firefighting appliances in operating condition, training and regular drills for employees, and use of alarm systems. Also, there should be prohibition of smoking in dangerous areas and use of open flames for carrying out gas cutting operations. There shall be good housekeeping as well as proper electrical lighting and wiring. The objective should be prevention, control, and extinguishing of fire.
- Starting of IC engines (cars/trucks) shall not be allowed in the vicinity of storage tanks for LPG, propane, diesel because sparks can come out from exhaust pipes of these vehicles (Table 2.1).
- Mechanical hazards: This refers to accidental injury to operating persons working in the vicinity of the machines. A well-designed machine must have safety guards, pressure release and tripping systems and audiovisual alarms for danger.
- Electrical hazards: Accidents occurring due to electrical hazards are:
  - Shock by AC current and burn by DC current due to poor protection from high voltage
  - Faulty or poor wiring
  - Fire from overload
  - Fire from capacitor discharge
  - Static electricity discharge
- Preventive measures from shocks and electrical fires include:
  - The cables shall be laid away from high temperature surfaces as far as possible.
  - They shall not be laid directly below piping carrying corrosive liquids (which can fall on them in case of a leak from piping above.)
  - Provide proper enclosures for high voltage equipment, capacitors, and joints.

Table 2.1 Classification of fire and corresponding extinguishing agents

Class of fire	Extinguishing agent
Wood, rags, paper plastics, etc.	Water, foam, dry chemical, automatic sprinkler system
Flammable vapours and liquid	Foam snuffing steam, carbon dioxide, dry chemicals. "Water fog or spray can be used to avoid spreading of wildfires"
Electrical fire	Dry chemical and carbon dioxide are useful, foam and water must not be used
Fires of combustible metals such as sodium and magnesium	Normal extinguishing agents are of no use, special types are required

  – Proper wiring and insulation for all equipment.
  – Regular inspection of the insulation, earth resistance, etc.
  – Proper controllers of high and low voltage to be handled by only trained, qualified persons who are licensed to work on high voltage equipment.

- Chemical (toxic) hazards: Chemical safety data sheets provide information on the safe handling of hazardous chemicals. Information about the important properties of materials being handled, operating conditions, safety data sheets; and drawings must be available for all equipment. There should be good local and general ventilation exhaust systems available to control such health hazards.
- Health hazards: The rules and regulations set by the state authorities for dealing with the health hazards and industrial hygiene as well as the specific instructions from local factory inspectors (statutory authorities) must be complied with. Occupational health hazards should be avoided by controlling dust, gases, vapors, X-rays and radioactive radiations.

2.5 HAZID: Hazards Identification

Many chemical plants operate with processes involving high pressure, high temperature, and have process equipment and machineries that handle corrosive, (highly) reactive, inflammable, or toxic (CRIT) substances; or can hold considerable quantities of such materials (*materials under process*) inside them.

Dangerous situations can develop during plant operations or even when the plant is not operating, e.g. failure of external cooling water spray on storage tanks that have highly volatile and inflammable solvents can be dangerous.

Hazards could be events that can

- Lead to injury to people, either inside or outside the plant.
- Harm or badly disturb the environment due to fire, release of toxic materials.
- Damage the assets in the plant (process units, machinery, piping). *This can result in loss of output of products (unable to meet commitment to clients regarding quantity, quality, or delivery schedules).*

The plant engineers and senior executives must examine the processes, equipment, operating conditions, and the procedures/methods that shall be followed when a new facility for manufacture is being set up. This can create awareness at the design stage and can assist in incorporating suitable changes in the process parameters, modifications in equipment and piping designs, and provision of appropriate safety systems (safety valves, rupture discs, and devices).

2.6   Identification of Hazards by Critical Examination of

- Process itself (expected operating conditions of pressure, temperature, flow rates, material and energy balance, and likely generation of effluents).
- Whether the process and technology is a proven one or is being implemented for the first time (just after conducting trial runs in a pilot plant).
- Piping and instrumentation diagrams P & ID.
- Effluent treatment plant and air pollution control facilities provided; safe disposal of hazardous waste produced, if any.
- Working flow sheet with detailed description of all operations and processes
- Storages of all toxic, inflammable, and corrosive materials shall be at isolated places. Check proper cordoning and dyke walls around the storages with provision of cooling water spray (if necessary for safe storage).

2.7   Check List for Prevention of Dangerous Situations

- Procurement of equipment and bought out components (valves, gaskets, instruments, internal fittings) is to be done from a reputed manufacturer/fabricator only.
- Only certified safety valves shall be fitted.
- The design, fabrication, and testing of plant equipment (which will be used to handle hazardous materials) must be done as per established codes and standards. Statutory rules and regulations must be complied with.
- Quality assurance tests must be carried out to ensure quality before the equipment are accepted for use.
- *Purchaser should not accept such equipment or machinery unless it meets all the above conditions.*
- Any shortcomings observed in the equipment must be rectified immediately to prevent accidents in future.
- A clause for Assistance/Supervision for Erection and Commissioning by the vendor must be included in the contract for the purchase of major equipment.
- Safe operating procedures, instructions for dos and don'ts are to be obtained from original equipment manufacturers (OEM) and discussed with technical consultants and plant engineers. These should be understood properly by all concerned.
- Structural design for stability of all the supporting structures must be obtained from certified and experienced structural design engineer. It must take into consideration the load bearing capacity of the land, local climatic conditions and provide adequate corrosion allowance since it is going to support equipment in a chemical plant. *Sufficient anti-corrosive treatment shall be carried out for all the structural members and all load bearing columns, beams, etc.*
- Safety margins for wind loads and earthquakes of moderate intensity (Richter scale of 8.0) should be used.

- Proper earth connections must be provided for all units, piping, etc., which shall handle inflammable, explosive, and toxic materials.
- Structural audit must be done by licensed and experienced engineers before placing any load of process units, machinery on them.
- Erection and commissioning procedures as recommended by vendors/experienced team of engineers shall be strictly followed.
- Environmental Impact Assessment shall be carried out and any other statutory instructions must be complied with.
- Permissions must be taken from statutory pollution control authorities for establishing and operating the plant by demonstrating the ETP and APC arrangements (including handling of sudden shock loads of effluents due to any reason).
- Arrangements must be done for safe disposal of hazardous wastes.
- Marketing: Safe procedure to be followed for sampling dangerous materials.
- Incoming tankers (raw materials and fuels) and product delivery tankers must be equipped with earth connection, safety valves, fire extinguishers, and any other device prescribed by law. *Guidance should be given to clients for safe unloading from tankers.*
- Hazardous operations should be isolated by location in separate buildings or isolated by brick walls/reinforced concrete walls (RCC) which can limit effect of hazards.

2.8   Ensuring Safety in Plant

The following mistakes/shortcomings should be avoided for safe working of the plant.

(Readers shall note that this is not an exhaustive list. More points can be added by experienced engineers)

- Excessive storage of dangerous materials CRIT (corrosive, reactive, inflammable, toxic) in premises; near hot surfaces of furnaces or ducts carrying flue gases.
- Non-compatible materials stored in close proximity of each other.
- Storage tanks not provided with dyke walls and warning alarms for overflow.
- Electrical cables laid directly below pipelines carrying acids, strong alkali or steam lines (any leak from pipelines can cause short circuits).
- Feeding of raw materials into process reactors by large capacity pumps instead of metering pumps of lesser capacity (just adequate for the operations).
- Standby power supply arrangements are not available.
- Standby cooling water supply arrangements are not provided.
- Insufficient lighting of working areas; near key equipment.
- Exhaust system/proper ventilation not provided for working areas.
- Narrow ladders or less number of independent escape routes from areas handling high pressure/inflammable/toxic materials

- HP steam line connected directly (without Pressure Reducing Station) to heating jackets of process vessels. The HP steam line must have a safety release valve.
- Heating jackets of process vessels not designed to withstand high pressure.
- Heating jackets of process vessels not provided with a separate safety valve.
- Earth connections not properly provided to storage/process tanks, process units, and pipelines handling inflammable materials.
- Flame proof motors, lighting not provided in areas handling inflammable materials.
- Ordinary shaft seals provided instead of mechanical seals for pumps handling dangerous liquids.
- Exceeding the safe operating limits (as per design/as recommended by OEM Original Equipment Manufacturer) for process equipment such as reactors, pressure vessels, etc.
- Boilers and pressure vessels: Necessary precautions, safety valves, and necessary fittings as instructed by Statutory Boiler Inspector, Factory Inspector are not complied with.

2.9 HAZOP (Hazard and Operability) Study

Hazard and operability study (HAZOP) is the systematic examination for identifying all plants and equipment hazards and operability problems of new grass roots plant or a planned increase in capacity, diversification to additional products, modification in the process, or operation of an existing plant.

The aim should be to identify hazards and to assess the hazard potential of operability problems or malfunctioning of individual items of equipment which can be dangerous to people, individual equipment, facilities in the premises, and external environment. The plant should be operated as per standard operating procedures evolved after detailed HAZOP studies. These procedures can be revised after the HAZOP studies are carried out in future again.

2.9.1 When Should HAZOP Study Be Carried Out?

Examine when the details of the process flow/P & ID are completed. It is to check any major omissions or significant features.

This study should generally not be carried out before completion of the P & ID.

HAZOP study of an existing plant: The P & ID shall be prepared as per actual built up plant before such study is taken up. Then include all proposed changes in equipment, piping, etc. and modify the P & ID accordingly.

Fresh HAZOP study shall then be carried out on this modified P &ID.

It should never be carried out on an incomplete or incorrect P & ID or if it is not up-to-date.

The working process flow sheet—which gives the details of operating procedures—should also be examined to ensure that all possibilities of mal-operation have been considered.

A large project has many sub-systems and hence the HAZOP study on such a project may take many weeks or months. In this case there could be two options:

a. Do not carry out the detailed design or proceed with construction of the plant until the HAZOP study is over. However, this will need holding on certain construction work and the project may get delayed. The advantage is that it can save the cost of corrections *to be carried out later on.*
b. The detailed design and plant construction may be continued with a calculated risk to make the changes *later on if necessary.* Though the project may not be delayed, these last minute changes may be not exactly as required by HAZOP study.

Purchaser shall discuss any new machinery being procured (or the changes being made in existing machinery) with shop floor Process and Maintenance personnel to ensure that they have fully understood the operation and maintenance of the new items. The likely adverse effect of any wrong operation, stoppage of utilities (power, cooling water, compressed air) or excess flow of reactants on the working personnel, other nearby units or surroundings shall be explained in detail to all concerned operating and maintenance personnel.

2.9.2   After Starting Operations (and Running the Plant for a Few Days/Weeks)

Certain undesirable changes may get introduced inadvertently in the operating conditions for the process parameters (e.g. temperatures, concentrations, feed rates to be maintained), plant equipment or the operating procedures as per shortcuts taken by operators for their convenience during course of operations. These shortcuts can be dangerous.

Some of these changes may not be compatible with the original design (as they tend to bypass certain safety checks and procedures) and *may result in operating the equipment or process units above their safe limits or may result in a dangerous situation unknowingly.*

Examples:

1. Operators may not check the alarms for high/low levels or pressures every shift.
2. Safety alarms or interconnections for tripping systems are bypassed sometimes.

It is therefore advisable that detailed studies HAZID and HAZOP shall be carried out every three to six months (but not later) after the plant has been commissioned or whenever new equipment/machinery are being procured.

These studies shall be carried out by the technical team (assigned for HAZOP analysis) by looking into the actual operating parameters and the operating procedures being followed.

They should also check the adequacy of the safety devices and recommendations for safe operations made in earlier studies and whether any further modifications or additional safety precautions have become necessary for further improvement.

2.9.3   Steps for Carrying Out HAZOP Study

Formation of a Technical Team
A technical team should be formed to carry out HAZOP study. It should include experienced designers, consultants, fabricators, plant engineers (chemical, mechanical, electrical, instrumentation, civil) as well as members from the erection, operating and maintenance departments. The team shall also include experts from Environment, Health and Safety department.

The technical team formed to carry out HAZOP must have a copy of the Material Safety Data Sheet (MSDS) of all substances which will be handled in the process and plant premises. All of them should study the MSDS before giving their opinions on the P&ID and suggestions for any changes to be done.

Members of the technical team and plant engineers should look in to the information which may be available on the link given here: https://en.m.wikipedia.org/wiki/NFPA_704. The FIRE Diamond should also be referred to while carrying out HAZOP (for handling various substances) in the plant.

The team should study the following:

* Every detail of the process design as well as P & ID.
* Design of process units, key equipment and machinery, and *suitability of their materials of construction for the chemicals to be handled.*
* Specifications of piping and instrumentation.
* Starting and operating procedures (*especially while operating at overload i.e. above rated capacity to meet urgent demand in the market*).
* Procedures for stopping the plant and carrying out maintenance in a safe manner.
* Safety valves, vents, and devices provided for personal and plant safety.
* Incorporation of additional safety and built-in precautionary measures (e.g. additional safety valves, rupture discs, electrical interconnections for tripping the concerned equipment drives, safety gas seals, smoke and gas detectors, warning alarms, fire extinguishers, and additional overhead water storage tanks for cooling).
* The possibilities of accidents in case of malfunctioning of new process units and machinery, wrong operations, choked piping, inoperative valves, reverse (or excess) flow of process streams, failure of cooling system, and excessive feeding of raw materials.

- What can happen if in case the plant personnel do not properly understand the operation or maintenance procedures.

The team should hold detailed discussions on the above matters and consider the possibility of adverse effect on plant personnel, machinery, and the environment (surroundings).

They should work out feasible preventive measures for ensuring safety.

The report of the discussions and recommendations shall be sent to senior management for approval and sanction of funds for implementation.

2.9.4   Suggested Procedure

The technical team shall carefully examine each equipment, piping, control instruments, and segments of the plant and identify all possible deviations from normal operating conditions.

This is to be done by fully defining the purpose of each segment and then applying guide words for each segment as follows:

1. No—No part of aim is achieved and nothing else occurs e.g. no flow
2. More—Quantity increases. e.g. higher temperature
3. Less—Quantity decreases e.g. low pressure
4. As well as—Qualitative modification/increase e.g. an impurity
5. Part of—Qualitative decrease e.g.: only one of two components in the mixture
6. Reverse—Opposite e.g. backflow
7. Other than—No part of aim achieved and something different occurs e.g. flow of wrong material

These guide words are to be applied to variables like flow, temperature, pressure, liquid level, and composition, which can affect the process. The results of these deviations are to be assessed and the necessary corrections are to be suggested.

Case Studies
Case Study 1: Hazop Study of Distillation Column
A typical distillation arrangement contains several major components:

- A vertical vessel where the separation of liquid components is carried out.
- Column internals such as trays/plates and/or packing which are used to enhance component separations.
- A reboiler to provide the necessary heat of vaporisation for the distillation process.
- A condenser to cool and condense the vapour leaving the top of the column.
- A reflux drum to hold the condensed vapour from the top of the column so that liquid (reflux) can be recycled back to the column.

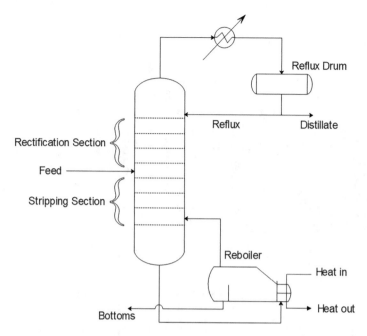

Fig. 2.1 A typical distillation column arrangement

The distillation column is heated by steam coils inside or by hot gases in the external reboiler. Incoming mixture of liquids is fed by centrifugal pump, and vapors are taken out at different temperatures (Fig. 2.1, Table 2.2).

Case Study 2: Hazop Study of Steam-Jacketed Reactor

A jacketed vessel is a container that is designed for heating its contents, by using a heating "jacket" around the vessel through which a heating fluid is circulated.

a. Conventional jacket (an external enclosure)
b. Limpet coiled jacket

Internal steam coils are also used for heating in certain designs.

Steam is admitted in the outer jacket/limpet coils at controlled pressure. External steam jackets may generally be less thermally efficient than internal submerged coils due to radiation losses to the surroundings.

Agitation can also be used in jacketed vessels to improve the homogeneity of the fluid properties such as temperature or concentration or to improve the rate of reaction.

An agitated steam-jacketed reactor is to be operated by feeding (liquid) raw materials (Fig. 2.2, Table 2.3).

Table 2.2 Hazop study of distillation column

Parameter	Guideword	Deviation	Possible Cause	Consequence	Action
Feed flow	No	No flow	• Feed pipe blockage • Feed control valve shut • Feed pump failure	• Column dry out • No operation	• Install low-level alarm • Attend Feed control valve • Start standby pump • Emergency plant shut down
	Less	Less flow	• Feed pipe blockage • Feed control valve not working properly • Check valves fail • Feed pump problem	• Column overheating • Column dry out • Changes in product quality	• Install low-level alarm • Attend control valve/check valve • Operate through bypass line • Emergency plant shut down • Attend feed pump
	More	More flow	• Control valve is fully opened	• Flooding in the column • Changes in product quality • Temperature decrease in column	• Install high-level alarm • Attend feed control valve
Steam supply Pressure to reboiler coils	High	High pressure	• Improper working of steam pressure regulator	• Risk of leak due to overpressure • Temperature increase of feed mixture in reboiler	• Install high-temperature alarm for reboiler • Install safety valve at steam coil inlet • Attend steam pressure regulator
	Low	Low pressure	• Improper working of steam pressure regulator	• Temperature decrease of feed mixture in reboiler • Process delayed	• Provide a bypass line for steam supply in exceptional case only • Attend steam pressure regulator
Condenser cooling water supply	Low	Low flow	Cooling water pump has stopped/valve in water line is closed	• Pressure build up in condenser due to less condensation • Production rate reduced	• Install low-water flow alarm • Open the valve in water line • Use standby pump for water supply

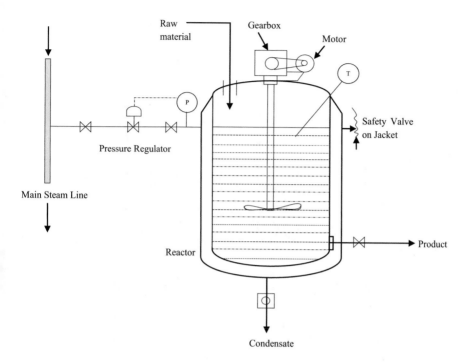

STEAM HEATED JACKETED REACTOR (WITH SEPARATE SAFETY VALVE)

Fig. 2.2 Schematic steam heated jacketed reactor (with separate safety valve)

Case Study 3: Hazop Study of Oil-Fired Furnace

Furnaces have many uses such as metal extraction from ore (smelting), in oil refineries, for manufacture of sulphuric acid, for steam generation, and for carrying out chemical reactions. Some common furnaces are the following:

a. Gas-fired furnaces
b. Oil-fired furnaces
c. Electrically heated furnaces
d. Coal-fired furnaces.

These furnaces operating at high temperatures are provided with heat-resistant refractory lining of fire bricks, castable refractory, insulating bricks, etc. to protect their shells and to minimise heat loss to surroundings.

A refractory lined furnace is provided with an oil firing system which consists of a day tank, oil supply pumps, oil burner, and air blower. The exit flue gases are to be used for heating a process reactor (Fig. 2.3).

The HAZOP study for this oil-fired refractory furnace will be as below.

Table 2.3 Hazop study of steam-jacketed reactor

Parameter	Guideword	Deviation	Possible cause	Consequence	Action
Reactant flow	No	No flow	• Feed pipe blockages • Feed control valve shut • Feed pump failure	• No operation • High temperature inside • Changes in product quality • Fall in level if the reactants evaporate	• Install low-level alarm • Install high-temperature alarm for inside material • Operate standby feed pump • Operate through bypass line for feed control valve • Emergency plant shut down
	Less	Less flow	• Feed pipe blockages • Feed control valve failure • Pump failure	• Changes in product quality • Delay in operation • Fall in level if the reactants evaporate	• Install low-level alarm • Install high-temperature alarm for inside material • Operate standby feed pump • Operate through bypass line for feed control valve
	More	More flow	• Auto-feed control valve is fully opened • Feed pump operated on high load	• Flooding in the reactor • Changes in product quality • Rise in level in reactor	• Install high-level alarm • Attend auto-feed control valve operation
Steam pressure for jacket	High	High pressure	• Steam pressure regulator not working properly • Operator is not attentive	• Damage to jacket due to overpressure • Explosion risk if steam reacts with reactants • Temperature increase of reaction mixture	• Check steam pressure regulator regularly • Install high-pressure alarm on jacket • Provide safety valves on jacket and the reactor vessel • Carry out regular pressure tests for the jacket
Steam pressure for jacket	Low	Low pressure	• Steam pressure regulator not working properly • Operator is not attentive • Low pressure in main line	• Temperature decrease of feed mixture • Process delayed	• Install low-pressure alarm • Check steam pressure regulator every shift • Check steam pressure in main supply line

Agitation	No	No agitation	• Agitator stopped working • Impeller dislocated from shaft • Foreign material have clogged the impeller	• Desired mixing efficiency not achieved • Delay in reaction • Loss in product yield and quality	• Monitor current drawn by agitator motor • Provide speed indicator at control panel for agitator shaft
	Low	Low agitation	• Agitator not working at desired speed • Foreign material have stuck to the impeller • High viscosity of reactants	• Delay in reaction • Loss in product yield and quality	• Speed of the agitator rotation to be monitored and maintained by controller • Monitor current drawn by agitator motor
	High	High agitation	• Sudden surge in voltage supplied to the motor • High speed setting for gear box • Very low level in reactor	• Increase in temperature of reaction mixture • Chance of explosion in reactor due to very fast reaction	• Speed of the agitator rotation to be monitored • Provide safety valve on reactor • Provide variable frequency drive to the motor • Check speed ratios of gear box pulleys
	Reverse	Reverse agitation	Wrong direction of rotation of motor (wrong electrical connection). This is likely when motor is replaced	• Desired mixing efficiency not achieved • Delay in reaction	• Proper direction of rotation of motor to be confirmed in the beginning itself • Monitor current drawn by motor

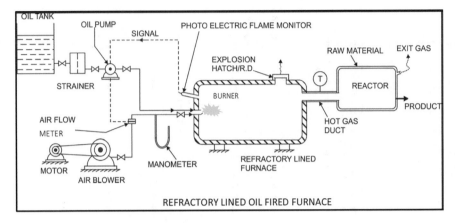

Fig. 2.3 Schematic of refractory lined furnace

Parameter 1: Fuel Oil Flow
A steady fuel oil flow is required to maintain the flame in the furnace and sustain its temperature. This steady flow can change to either an increased or decreased flow rate due to various reasons as given below and can have undesired consequences.

a. Clogged oil lines/strainer in oil line
b. Control valve stuck in closed position
c. Tripping of the motor for oil pump
d. Malfunctioning of oil pump

This can lead to possible extinguishing of the flame and a decrease in the furnace temperature, which can adversely affect product quality. It can be prevented by proper checking of the oil feeding system and installing an alarm for low oil flow.

It will be useful to provide a photoelectric flame monitor with an audio-visual alarm in the control room to indicate extinguishing of the flame.

The opposite can occur in case there is a sudden increase in the fuel oil flow rate. This can occur if the control valve for oil flow fails and gets stuck in an open position, thus failing to prevent an oil surge, which could be caused by excess feeding of oil by the pump. This can lead to a sharp rise in the furnace temperature and can even lead to an explosion. Provide a rupture disc/explosion hatch (with a gas release duct) on the furnace shell to release the pressure.

It can also cause hot spots in the furnace shell. Monitor the temperature of the shell at various spots regularly.

The adverse effect on product quality can be prevented by following proper maintenance of control valve for oil flow and by providing a metering pump (of proper capacity) for controlled feeding of oil. There shall be a locking arrangement on the flow setting of the oil pump. A high oil flow alarm may be provided as an additional precaution.

Parameter 2: Air Flow

A steady air flow along with the fuel oil is required for the furnace for combustion of oil by sustaining the flame. This air flow can get hampered or completely stopped due to blockages in the air line. It can happen if the valves in the air line are shut or stuck in closed position.

This will lead to a reduction in the amount of air supplied to the furnace and can extinguish the flame. Low air flow alarms should be installed to prevent this from occurring. Air flow rate can be monitored and maintained using a control system. The dampers/control valves in the air lines shall be maintained properly.

Sufficient air blower pressure is required to maintain the air flow to the furnace. This pressure can reduce due to various causes such as:

- Leakage in the airline pipes
- Malfunctioning of control dampers in air lines.
- Air blower motor trips or the drive belts are slipping
- Clogged strainer screen at suction side of the air blower. Should be cleaned every day.

Preventive Arrangements

- Installation of low-pressure alarms in the air line to the furnace
- Continuously monitoring of the current drawn by the air blower motor
- Keeping the air blower drive belts in proper working condition
- Blower speed indicator may be provided in control room (Table 2.4).

Case Study 4: Typical Scheme for Production of Liquid SO_3

30% Oleum is boiled in a steam-heated boiler to generate SO_3 vapours which are condensed by a water-cooled condenser to produce liquid SO_3.

Flow of 30% oleum to the boiler is carefully controlled. It is continuously monitored by a rotameter. Temperatures of SO_3 vapours and exiting depleted oleum stream from the boiler are also monitored (Fig. 2.4, Table 2.5).

Notes

- In a variant of the system, hot process gases are used for boiling oleum in the boiler instead of steam. This eliminates the risk of violent reaction due to ingress of steam in the oleum.
- A special design is required for the condenser to avoid cooling water flow under pressure. Trickling film of water is to be used. Any tube leak can result in SO_3 vapours entering the water side (instead of water entering into SO_3 side).
- A conductivity sensor/indicator in the exit water stream from the condenser with an alarm will give a warning to the plant operator about a tube leak.

Table 2.4 Hazop study of oil-fired furnace

Parameter	Guideword	Deviation	Possible cause	Consequence	Action
Fuel oil flow	No	No flow	• Oil feed pipe clogged • Strainer in oil line is clogged • Oil feed controller valve stuck in closed position • Feed pump motor tripped • Feed pump malfunction • Burner check valve stuck in closed position	• Flame extinguished • Sharp drop in temperature of flue gases • Heating of reactor will stop • Reactor working can get disturbed	• Install low-oil flow alarm • Attend oil feed controller valve • Clean strainer/oil feed pipe • Start standby pump • Install photoelectric flame monitor with audio visual signal • Attend burner check valve • Emergency process shut down
	Less	Less flow	• Feed pipe or strainer clogged • Oil feed controller valve not working properly • Burner check valve is almost closed • Feed pump malfunction	• Low furnace (exit) flue gas temperature • Low reactor temperature • Changes in product quality due to low reactor temperature	• Monitor flue gas temperature • Clean oil feed pipe and strainer • Attend oil feed controller valve • Open burner check valve properly • Attend feed pump
	More	More flow	• Malfunction of oil feed controller valve (opened too much) • Burner check valve (opened too much)	• Excess unburnt oil in the furnace can cause explosion • Changes in product quality • Temperature of flue gases can suddenly increase • Hot spots on furnace shell	• Install rupture disc RD/explosion hatch on the furnace • Provide high flow limiting stoppers on feed controller valve • Attend oil feed controller valve • Adjust oil flow by burner check valve also
Air supply to the furnace for oil firing	High	High air pressure/high air volume	• Improper working of control valve in air line • Valves in air lines opened too much	• Fall in temperature of flue gases • Burner flame may flicker or extinguish	• Install photoelectric flame monitor with audio visual signal • Monitor air flow and air pressure to furnace
	Low	Low air pressure/low air flow	• Improper working of control valve in air line • Valves in air line almost closed • Blower suction screen clogged	• Burner flame may flicker or extinguish • Production rate reduced	• Attend control valve in air line • Monitor burner flame • Install blower speed monitor if possible • Check valves in air line • Clean blower suction screen
	Very low/stopped	Very low air pressure/low air flow	• Problem with coupling or belt drive • Incorrect motor provided to air blower (low speed, low HP) • Wrong direction of rotation of	• Unburnt oil vapour in the furnace can cause explosion • Production rate reduced	• Provide low air flow/pressure alarm • Attend coupling/belt drive • Check motor speed, and current drawn • Check direction of rotation of blower

References/for Further Reading

1. "Process Equipment Procurement in the Chemical and Related Industries", Kiran Golwalkar, Published by Springer International Publishing, Switzerland.
2. "Production Management of Chemical Industries", Kiran Golwalkar, Published by Springer International Publishing, Switzerland.
3. "Chemical Process Safety: Fundamentals with Applications", Daniel A. Crowl and Joseph F. Louvar, Second Edition. Pearson Publisher, 2001

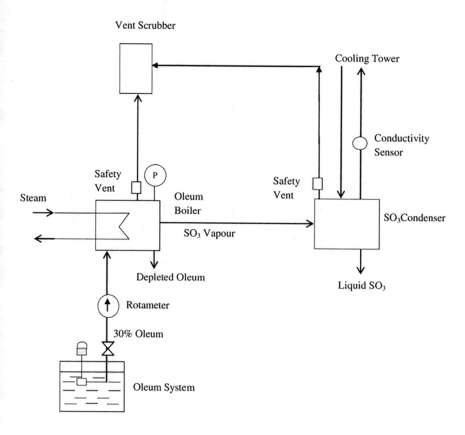

Production of Liquid SO_3

Fig. 2.4 Typical scheme for production of liquid SO_3

Table 2.5 HAZOP study for scheme in Fig. 2.4

Parameter	Guideword	Deviation	Possible cause	Consequence	Action
Steam supply	High flow	High pressure	• Steam pressure regulator malfunctioning	• Excessive generation of SO_3 vapours • High pressure either in condenser or in boiler or both • Explosion due to boiler tube leak (steam ingress in oleum)	• Provide steam pressure regulator with setting on lower pressure • Provide safety valves/release vent on boiler and condenser both. Connect the vent lines to scrubber • Design, fabricate, and test the oleum boiler and SO3 condenser as if they are pressure vessels
	No flow	Low pressure	• Steam stop valve not opened • Pressure regulator not working • Steam trap not working	• Reduction in generation of SO_3 vapours	• Open steam stop valve • Check setting of steam pressure regulator • Attend steam trap
Cooling water to condenser	No flow	• No condensation of SO_3	• Valve in water line not opened • Cooling water pump not working	• High pressure in condenser or boiler or both	• Provide alarm for no flow of water at inlet of condenser • Provide safety valves/release vent on boiler and condenser both • Open valve in water line • Attend water pump • Start standby water pump
Oleum flow at inlet to boiler	No low/low flow	• No oleum in the boiler • Very little oleum in boiler	• Valves in oleum feed line not opened/only partially opened • Oleum pump improper working	• No production of SO_3 vapours • Very low production of SO_3 vapours	• Open valves in feed line carefully and adjust the flow on rotameter • Check oleum pump
	High flow		• Valves in oleum feed line opened too much	• High vapour pressure in boiler • Liquid oleum entry in SO_3 line	• Provide bigger diameter pipe for exit oleum (return) stream from boiler • Provide safety valve/vent for boiler. Connect vent line to scrubber • Provide an orifice in feed line to boiler to limit excess flow • Provide high-level alarm for boiler. • Monitor rotameter readings

Chapter 3
Materials of Construction

Abstract It is essential to select a proper material of construction for fabrication of the process units and piping; and for the machinery to be used. Operating conditions and properties of all substances which shall be handled are to be examined in detail. Selection of proper MOC shall be confirmed after study of the properties of the MOC.

Appropriate refractory linings/corrosion-resistant linings and coatings shall be provided to the equipment to protect them from severe operating conditions of high temperature, corrosive conditions, etc. Galvanic corrosion shall be controlled by sacrificial anodic or impressed cathodic protection to the equipment and structures.

Keywords Basis for selection of material of construction · Physical properties (volumetric, thermal, and mechanical) · Ferrous metals · Stainless steels · Nonferrous metals · Nonmetals · MOC for high temperature and low temperature service · Corrosion · Electrochemical · Impressed current cathodic protection · Anodic protection · Corrosion inhibitors · Refractory lining · Acid-resistant bricks lining · Cladding · Corrosion-resistant materials · Coatings and paints

3.1 Introduction

Proper selection of material of construction helps in minimizing risk of breakdowns, contamination of products and development of unsafe conditions due to corrosion.

Common materials of construction used for process plants include carbon steel, stainless steel, steel alloys, graphite, glass, titanium, plastic and many more. Alloying the steel is one of the most effective methods wherein the properties of various metals are used to provide added corrosion resistance to the resultant product alloy. Typically, nickel, molybdenum and chromium are added to steel to make a corrosion-resistant alloy.

However, alloy steels are expensive as compared to mild steel.

Materials can be selected depending on operating conditions and properties of the materials being handled in the process plant. Options are limited for highly corrosive media.

Design of the process equipment is to be done after selection of appropriate material of construction.

3.2   Basis for Selection of Material of Construction in Process Industries

- Corrosion resistance for process conditions such as temperature, pH, nature of fluid handled, and presence of trace impurities.
- Mechanical strength i.e. tensile strength, toughness, hardness and wear resistance, creep, and fatigue resistance requirements for process conditions.
- Estimation of cost of the equipment (capital expenditures) to ensure financial limits are not exceeded.
- Compatibility of material of construction with plant operating conditions.
- Commercial availability.
- Ease of maintenance (repairs)/replacement of the process unit.
- Expected operating life with the MOC—whether the operation is continuous or batch process.
- A test piece of the material to be used for construction should be tested in actual working conditions (or simulated in approved laboratory) and the properties shall be checked before making final choice.

Physical Properties
The properties of materials should be known in order to evaluate the behaviour of the material in the media (*which they will be subjected to*) for making a proper choice. However, the choice of material of construction also depends upon the ease of fabrication and cost. For the ease of fabrication, properties such as machinability, weldability, and malleability should be considered.

Physical properties include volumetric, thermal, and mechanical properties.

Volumetric Properties Volumetric properties are related to the volume of solids such as density which is an important consideration in material selection for a given application. Strength is also important, and the two properties are often related in a strength-to-weight ratio, which is tensile strength divided by its density. Strength-to-weight ratio is useful in comparing materials for applications such as agitators, mixers, and ribbon blenders.

Mechanical Properties The mechanical properties of a material are those which affect the mechanical strength and ability of a material to be fabricated in suitable shape. Some of the typical mechanical properties of a material include strength, toughness, hardness, hardenability, brittleness, malleability, ductility, creep, resilience, and fatigue.

Thermal Properties Thermal properties include thermal expansion, melting point, specific heat, thermal conductivity and heat of fusion. Thermal properties should be considered because heat generation is common in many processes while in some cases heat is to be provided to carry out the process. Good thermal conductivity is necessary for efficient heat transfer.

Thermal expansion is used in shrink fit and expansion fit assemblies. In this method the part is heated to increase size or cooled to decrease size to permit insertion into another part. When part returns to ambient temperature, a tightly fitted assembly is obtained.

Materials having a high melting point can be considered for construction of process units operating at higher temperatures.

Important Consideration Welding during fabrication can cause thermal stresses due to thermal expansion which need to be relieved by heat treatment.

3.3   Common Materials of Construction Used in Process Industries

(Please refer to Appendix A Properties of Materials of Constructions)

- Metals
- Ferrous metals: Various types of steel depending on percentage of carbon along with small percentages of other metals such as nickel, chromium, vanadium, manganese, molybdenum, tungsten added to iron for alloying purpose.
- Non-ferrous: copper, nickel, aluminium, lead, etc. and their alloys.
- Non-metals:
- Plastics, glass, rubber, graphite, concrete.

3.3.1   Ferrous Metals

Ferrous metals are more commonly used as they are easily available and have good strength, ease of fabrication, machining, and forming properties. The properties can be modified by making ferrous alloys or steels.

Steel Steel is basically an alloy of iron and carbon with a small proportion of other metals such as nickel, chromium, aluminium, cobalt, molybdenum, and tungsten, etc. The carbon content in steel can range from 0.05 to 2.0%. Various types of steel are produced according to the properties required for their application by alloying iron with a small percentage of different metals.

Classification of steels based on their chemical compositions:

- Plain carbon steel: This has 0.05–0.6% carbon and very small quantities of manganese, silicon, phosphorus, and sulphur. These are further classified as low, medium and high carbon steel depending on percentage of carbon. Carbon steel is most widely used at ordinary to moderately elevated temperatures and non-corrosive services across many industries due to its strength, good ductility, and ease of workability. Carbon steel is relatively inexpensive as it contains relatively few alloying elements in low concentrations.
- Alloy steel: Lack of alloying elements makes carbon steel less resistant to the accompanying stresses particularly at extreme temperature or high pressure. Number of elements such as nickel, chromium, vanadium, manganese, molybdenum, and tungsten are therefore added to steel to improve its properties. Alloy steel is often categorized as low alloy steel, high alloy steel, and stainless steel based on the type of alloying materials.

3.3.1.1 Properties of Steel Are Improved by Alloying with Elements Such As

Aluminium: Aluminising of steel by hot dipping improves corrosion resistance and heat transfer. Hence it is used for tubes for gas-to-gas heat exchangers in sulphuric acid plants.

Chromium: increases strength, corrosion, wear, and abrasion resistance. Alloy steel with chromium is used in oil and gas industry.

Copper: improves corrosion resistance. It is added in small amounts (0.1–0.4 wt%).

Manganese: increases toughness, ductility, wear resistance at high temperature, and tensile strength of steel.

Molybdenum: improves machinability, increases high temperature strength and resistance to corrosion. Alloy steel with molybdenum is used in oil and gas industry.

Nickel: improves strength, corrosion, and high temperature resistance.

Titanium: improves hardness and strength.

Tungsten: improves hardness and strength at high temperatures.

Vanadium: increases toughness, strength, and corrosion resistance.

3.3.1.2 Composition, Properties, and Application of Ferrous Materials

Wrought Iron
This has low carbon and manganese. It is resistant to corrosion from alkali, concentrated acids, and organic compounds. It is used for elevated temperature applications, steam lines, heating coils.

Cast Iron

This has 2–4 wt% carbon and small quantities of silicon, nickel, chromium, and molybdenum. It has corrosion resistance to alkali solutions of less than 30% concentration, 98% sulphuric acid at 70–80 °C. It is used for pumps, valves, acid coolers, acid concentration vessels, machinery base plates.

Plain Carbon Steels

These have carbon content of 0.04–2.0% and small quantities of manganese, phosphorus, silicon. Depending on the carbon content these are further categorized as:

Low or mild carbon steel

- Have 0.04–0.3% wt carbon; low corrosion resistance.
- Easily rolled, forged and drawn, welded, and machined.
- Used for pipelines, fittings, pressure vessels.

Medium carbon steel

- Have 0.31–0.6% carbon; 0.08–1.70% manganese.
- Used for shafts, gears, structural support beams due to better strength.

High carbon steel

- The carbon content is 0.61–2.0%.
- These steels are harder and stronger than low- and medium carbon steels; and are difficult to bend, weld, or cut.
- Used for cutting tools and dies.

Alloy Steels

Low alloy steel has nickel content less than 10%; while chromium, manganese and molybednum are in small amounts; the carbon content is also low (0.2%).

These steels have high strength, toughness, and good corrosion and abrasion resistance; Presence of Cr, Mo, Ni in alloy steels improves strength and resistance to scaling, oxidation, with improved mechanical properties at high temperatures. They also have good weldability which leads to ease of fabrication. These are used for fabrication of pressure vessels

High alloy steel: These have 0.04–0.25 wt% carbon.

Stainless steel has 18–25 wt% chromium and 8–20 wt% nickel

They have high corrosion- and heat-resistance properties, increased strength, ductility, and toughness. The machinability and weldability are good. These are useful for fabricating process vessels.

3.3.2 Stainless Steels

Stainless steels have excellent corrosion- and heat-resistance properties. They have alloying elements such as chromium and nickel apart from small amounts of other elements such as manganese, silicon, nickel, and molybdenum.

The system of designation numbers for stainless steels by American Iron and Steel Institute (AISI) is widely used. Information is provided here for some of the common stainless steels. The readers can get more details from AISI.

Based on their crystalline structure and alloy composition, stainless steels are divided into three groups as given below.

Martensitic Stainless Steel These have up to 18% chromium, 2.5% nickel, 0.12% carbon, and a little molybdenum. They possess toughness, ductility, and mild / moderate corrosion resistance. Generally used for valve parts, pumps, and for handling water; and in oil and gas industries under mild corrosion conditions. The AISI grades are AISI 410, 414 (with nickel for better corrosion resistance), 420, 431 (chromium for corrosion resistance) 440 (increased corrosion resistance). Samples of the material shall be tested in the actual process conditions for corrosion resistance before the equipment fabricated from it are accepted for use.

Ferritic Stainless Steels These have a chromium content of 11–25% while the carbon content is low (less than 0.1%) and have good resistance to stress corrosion cracking; high temperature resistance to corrosion and oxidation. Some of the AISI grades are 405, 410, 430, 442, and 446.

They are used in exhaust systems, heat exchangers, furnace parts, and food processing equipment.

Austenitic Steels The chromium content is 18–25%; and nickel content 8–20%. These have good strength, ductility, and are highly rust-resistant. However, samples should be tested before use for corrosion resistance in case the process streams contain chlorides. They are nonmagnetic, have good strength and ductility, and are suited for high temperature services.

- Some of the AISI grades are AISI 301, 304, 304 L, 310, 316, and 316L. The carbon content is less than 0.10% generally; while Si is less than 1.0%.
- The letter L indicates a carbon content less than 0.03%.
- The grade 321 contains a small amount of titanium.
- The H grade has carbon content in range 0.04–0.1% and used for high-temperature applications such as in heat exchangers, furnaces, and exhaust systems.
- Used in chemical industries in food processing equipment and process piping.

Duplex and Super Duplex Stainless Steels These steels have higher amounts of chromium and molybdenum. These have very good corrosion resistance and strength and hence are used in oil and gas industries for piping, valves, pressure vessels, etc.

Alloy 20 is an austenitic stainless steel used in sulfuric acid and oleum service due to its very good corrosion resistance. It is also used in the petrochemical and plastics industries. Alloy 20 is used for circulation pumps in sulphuric acid and oleum plants.

It has the following typical alloying elements.

- Nickel (not less than 30%)
- Chromium (about 20%)
- Carbon, 0.06% maximum
- Copper, 3–3.5%
- Molybdenum, 2–2.8%
- Manganese about 2%
- Silicon not more than 1.0%
- Niobium in traces

Test certificate shall be obtained from the manufacturer of equipment (regarding the material used for fabrication) for the composition and corrosion resistance in the working conditions. A test piece may be tested in approved laboratory (Tables 3.1 and 3.2).

3.3.3 Non-ferrous Metals

Aluminium and Aluminium Alloys
Aluminium is light and easy to fabricate. Pure aluminium has low strength and is soft and ductile. It has high resistance to corrosion due to formation of protective oxide film on its surface. Aluminium alloys have better mechanical properties, high corrosion resistance, high thermal conductivity, non-toxic, non-magnetic, and good reflectors of radiant heat, can be used for low temperature service. They are used in dairy, food processing, and refrigeration industry. Commercial aluminium alloys such as Duralumin (4% Copper, 0.6% magnesium, 0.5% manganese), aluminium-zinc-magnesium alloy, and aluminium-silicon alloy have better mechanical properties and are used in structural components and fabrication of process vessels.

Aluminised steel tubes have good corrosion resistance and hence are used for gas-to-gas heat exchangers in sulphuric acid plants.

Copper and Copper Alloys
These have good ductility, malleability, high electrical and thermal conductivity, fair mechanical strength, and good ease of fabrication. They possess good corrosion resistance to strong alkalis, organic solvents, and sea water. They are used in food processing plants, manufacture, and recovery of organic solvents. Copper should not be used for oxidizing acids, ammonia, and carbon dioxide solutions. The following copper alloys are widely used:

Brass: Copper-zinc alloy with 45% zinc is used for making tubes and tube sheets for heat exchangers.

Bronze: Copper-tin alloy with 5–10% tin is used for making pumps, valves, and pipe fittings.

Copper-aluminium alloy has up to 14% aluminium and known as aluminium bronze. They have high resistance to oxidation and scaling and used for moderately elevated temperature services, making pump casings, valve seats, and condenser tubes.

Table 3.1 Grading system for different steels

Numbering system	Type of steel
1XXX	Carbon steels
2XXX	Nickel steels
3XXX	Nickel-chromium steels
4XXX	Molybdenum steels
5XXX	Chromium steels
6XXX	Chromium-vanadium steels
7XXX	Tungsten-chromium steels
8XXX	Nickel-chromium-molybdenum steels
9XXX	Silicon-manganese steels

Copper-nickel alloys have 10–30% nickel. They have good strength, corrosion resistance, and ductility and are used in steam condensers operating with sea water or with polluted cooling water.

Nickel and Nickel Alloys

Nickel has high corrosion resistance to fluids at low or elevated temperature. Following nickel alloys are used in chemical industries:

- Monel: Nickel-copper alloy (67% nickel and 30% copper) has high strength, corrosion resistance, and ductility. It is widely used in organic and food industries, in evaporators, and heat exchangers.
- Monel 400, Monel K-500—A nickel-copper alloy that is resistant to sea water/ phosphoric acid/salt solutions.
- Inconel Alloys: These are nickel, chromium, iron, manganese, and molybdenum alloys. They have good strength and toughness at both high and low temperatures and good corrosion resistance. Used in oil and gas industries, equipment handling alkalies.
- Hastelloy C-276—A nickel-molybdenum-chromium alloy that has excellent corrosion resistance in a wide range of severe environments. Used for Plate type heat exchangers for cooling concentrated sulphuric acid.
- Incoloy alloys: These are nickel, chromium, iron, and manganese alloys with small amounts of copper and titanium. They have good corrosion resistance.

Lead

This is soft and malleable and hence used for lining or cladding purpose. It has low creep and fatigue resistance. Usage of lead is avoided nowadays due to its adverse health effects.

Titanium

This combines high strength with low density to provide good corrosion resistance to sea water, oxidizing agents, and many organic acids across a range of temperatures. Due to formation of thin oxide film they have good corrosion-resistance properties. This can be used for pumps, valves, and heat exchangers.

Table 3.2 Materials of construction for typical process units

Sr No.	Equipment	Material of Construction	Remarks
1	Process reactor shell		
	(1) Corrosive media	SS 316, SS 316 L	MOC sample should be tested in laboratory
	(2) Non-corrosive media	M S, low alloy steel, SS 304	
2	Process reactor jacket	M S, SS 304, SS 316	Shall be pressure tested if heated by steam
3	External (limpet)heating coil	MS. SS 304, S S 316,	Shall be pressure tested if heated by steam
4	Melter internal coils	MS Sch 80; SS 316	Check corrosion resistance and pressure test
5	Dissolver		
	(1) Shell	MS; SS-316	With acid resistant lining (for corrosive media)
	(2) Agitator shaft	EN 8/EN 24	As per viscosity and density of medium, RPM
	(3) Agitator blades	SS 316, MS; SS 410	As per operating conditions pH, temp, viscosity etc.
6	Sight glasses	Toughened glass	Shall be pressure tested
7	Light glasses	Toughened glass	Shall be pressure tested
8	Distillation column	MS; SS-304, SS-316	
9	Condenser		
	(1) Shell	MS/SS-316	
	(2) Tubes	MS class C/SS-316	Example: BS 3059 CDS tubes for SO_3 condenser
	(3) Tube sheets	MS/SS 316	
10	Pan evaporator		
	(1) Shell	Internal acid resistant tiles	For evaporation of alum solution
	(2) Evaporator steam coils	Antimonial lead /lead lined MS	Evaporation of alum solution/ optional SS 316
11	Heat exchanger		
	(1) Shell	Aluminised steel shell	For corrosive gases
	(2) Tubes	Aluminised steel tubes	For corrosive gases
	(3) Tube sheets	MS, SS with refractory/ AR layer	To protect tube sheet and welding with tubes
12	(1) High temp. Heat exchanger	Carbon steel, SS-316, 316 L	For shell and tubes
	(2) High temp Exgr tube sheets	SS 316, SS 316 L, duplex steel	
13	Evaporator shell	MS with acid resistant lining	For corrosive service

(continued)

Table 3.2 (continued)

Sr No.	Equipment	Material of Construction	Remarks
14	Evaporator tubes	High chromium SS/ graphite	Check suitability for working conditions
15	Pumps for corrosive service	SS-316, Alloy 20, rubber lining	For wetted parts like impeller, shaft, volute
16	Pumps in non-corrosive service	CI, MS	CI for impeller/MS for shaft
17	Valves in liquid lines		
	(1) In corrosive service	SS 316/MS with teflon lining	PP/HDPE—check the temp and conc. of acids
	(2) In non-corrosive service	PVC, MS, Galvanised iron	
18	Butterfly valves in gas lines	Flap, shaft, bolts made of SS316	For corrosive gases
19	(1) Piping for process water	MS; GI, HDPE, PP	
	(2) pipes for dilute acid/ alkali	HDPE, Polypropylene	
20	Gas ducts	MS; SS-316	Internal ceramic paper lining for high temp.
21	Vessels for dil acid/ alkali	HDPE/polypropylene/ FRP	Generally for temperatures below 60 °C
22	Catalyst support trays	Cast Iron / SS 316	For converter in sulphuric acid plant
23	Steam lines	ASME B A-106	To be approved by boiler inspector
24	Condensate piping	MS Class C; A-106	

Note: Gas ducting after scrubbers and Induced draft fans for air pollution control are subjected to moist acidic gases. Hence these shall be lined by rubber/FRP as a protection against corrosion

Zirconium

Zirconium has good corrosion resistance to most organic and inorganic acids such as hydrochloric acid, nitric acid, salt solutions, strong alkalis, and some molten salts. It is used to make furnace linings, abrasives, and in glass and ceramic industries.

Tantalum

This offers very good resistance to corrosive/aqueous solutions. It can be used to repair glass lining of process vessels (which should be done under expert supervision by the original equipment manufacturer).

3.3.4   Non-metals

Plastics

Plastics have low weight, good thermal and electrical insulation, and corrosion resistance to weak mineral acids. Thermosetting and thermoplastics are the two main types of plastics.

A thermosetting plastic is a polymer that becomes irreversibly rigid when heated. These cannot be re-moulded after the initial forming. These are hard and brittle. Examples of thermosetting plastics are epoxies, phenolics, and polysters.

Epoxies are used for fibre composites, phenolics for adhesives, and polyester as a reinforced material along with glass are used for fabricating pipes, pipe fittings, vessels and tanks, and as lining materials.

Thermoplastics soften under application of heat and pressure and can be moulded and extruded to various shapes and sizes. These are available in the form of sheets, rods, and films. Common thermoplastics are fluorocarbons (polytetrafluoroethylene teflon), acrlylics, nylon, polyethylenes, polypropylene, polyster, and polyvinyl chloride.

Thermoplastics are used to make reaction vessels, storage tanks, ducting and ventilating systems, pipes and pipe fittings, gaskets, valves, piping, lining of process vessels, cooling tower, and absorber fill materials.

Composites

In composites, polymer matrix is strengthened by metallic or nonmetallic fibres. Composites have higher strength than the matrix alone.

Fiber Reinforced Polymers (FRP) has thermosetting polymers such as vinyl ester and epoxy for high corrosion resistance.

For intermediate chemical resistance, resins such as polyester and polyurethane are used. Graphite Fiber Reinforced Polymer (GFRP) is used for making automotive parts and in chemical plants.

Ceramics

They are inorganic nonmetallic solids with high corrosion resistance and also good thermal and electrical insulation properties. Ceramics are made from clay, quartz, silica, and feldspar.

Typical uses are: Silica glasses used for high temperature applications, aluminium nitride for making heat sinks, silicon for semiconductors, and silicon carbides for high temperature equipment.

3.4   Materials of Construction for Extreme Services

Selection of the materials, special alloys, etc. for the construction of process units shall be considered after testing (if possible) their samples in the worst working conditions or in an approved laboratory. This is to ensure reasonable service life.

However, any practical difficulty for fabrication of the given process unit should be discussed between the designer and fabricator.

There shall be commercial availability for fabrication in reasonable time (so that equipment delivery is not delayed).

Estimated cost of the unit should also be looked into.

3.4.1 Materials of Construction for High Temperature Service

The allowable stress of material of construction is temperature dependent as the tensile strength and elastic modulus of metals decreases with increasing temperature. Creep resistance, tensile properties and resistance to corrosion at elevated temperature and scaling are important properties that need to be considered for high temperature services. Materials which retain these properties at elevated temperatures must be selected. Low creep strength, tendency to decarburize and low corrosion resistance of plain carbon steel at temperatures above 400 °C make it difficult to use such steels at elevated temperatures. Addition of alloying metals such as chromium, molybdenum, and vanadium improve the strength and corrosion resistance of the material of construction.

However specific environments such as presence of hydrogen, sulfur, and sulfur compounds can cause degradation of mechanical properties such as embrittlement and corrosion at high temperatures and pressures.

Special alloys, such as Inconel are used for high temperature equipment such as furnace tubes and furnace parts. 25Cr-12Ni and 25Cr-20Ni austenitic stainless steels provide high creep strength and corrosion resistance for higher temperatures. Their use for pressure vessels/pressure piping or critical applications at high temperatures shall be only after proper testing of the material.

Austenitic stainless steel may be used for hydrogen services and chromium, molybdenum alloy steels for protection from attack by sulfur and sulfur compounds.

3.4.2 Materials of Construction for Low Temperature Service

Properties such as notch toughness (ability to absorb energy and deform plastically at stress concentrations) and failure due to brittle fracture need to be considered for selection of material of construction for low temperature service. Metals that are normally ductile can fail in a brittle manner at low temperatures.

Ferritic steels cannot be used for sub-zero temperatures due to reduced notch toughness at low temperatures. Many non-ferrous alloys such as aluminium alloys and austenitic steels can be used in cryogenic plants and liquefied-gas storages.

3.5 Corrosion

A major problem in process industry is corrosion of vessels, pipes, and valves. The loss of metals at the surface due to corrosion occurs mainly because free metals are in higher energy state due to which they tend to react with their environment. This produces more stable compounds such as metal oxides, metal sulfides, or metal hydroxides and leads to reduction in thickness of the metal.

The rate of corrosion depends on the material of construction, nature of fluids, process conditions such as fluid temperature, pH, and surrounding environment.

Corrosion is expressed as loss (weight or thickness) per unit time. Corrosion rates are typically measured in mils per year (1 mil = 0.001 in.).

A rough estimate of the expected service life of a process unit can be done by dividing the wall thickness of the unit by the corrosion rate. The unit may have to be discarded (as it can become too thin for safe use) *after reaching the end of expected service life.*

Corrosion tests should be carried out on a test piece of the material of construction to determine the loss after exposure of the metal to the corrosive environment. A corrosion allowance shall be thereafter added to the calculated required thickness of process vessels for increasing their useful life.

In petroleum industries corrosion caused by presence of hydrogen is very common. Hydrogen attack can lead to hydrogen blistering, hydrogen embrittlement, and decarburization which results in loss of ductility and tensile strength. Decarburization is removal of carbon in the steel. Hydrogen present in the surrounding media (gases) can react above 220 °C with carbides in the alloys generating methane gas (CH_4) which gets trapped into metal. As more hydrogen reacts, the concentration and pressure of methane gas increases. The increasing pressure of methane begins to cause fissures and then cracks in the grain boundary. At the same time, the loss of carbides due to reaction with hydrogen lowers the strength of the metal. At high temperatures hydrogen reacts with components of an alloy resulting in loss of strength. Hence materials of construction for use *in presence of hydrogen* are to be selected only after considering their resistance to attack by hydrogen at high temperature.

3.5.1 Harmful Effects of Corrosion

The metals that have leached out due to corrosion can contaminate the reaction media and produce undesired process reactions. These can affect product quality and yield; thereby lowering efficiency of the process. In certain cases, runaway reactions may also take place leading to fire hazard or generation of toxic products.

The reduction in thickness of process vessels can create unsafe conditions; increase cost of maintenance and may even need replacement of the corroded equipment.

3.5.2   Types of Corrosion

Dry or chemical corrosion: In this type of corrosion gases like hydrogen, oxygen, sulphur dioxide, halogens, and inorganic acid vapors chemically attack metal exposed to these gases. Rusting of iron due to formation of ferric oxide on the surface, formation of green film of basic copper carbonate on copper surface when exposed to moist surface containing carbon dioxide are examples of dry corrosion.

Wet or electrochemical corrosion: Electrochemical corrosion occurs when dissimilar metals or alloys are in contact with an aqueous solution (such as acid, base or salt) which acts as conducting liquid for transfer of electrons between the two metals. Under wet or moist conditions due to electrochemical reaction, the dissimilar metals act as anode and cathode and current flows between them through the conducting solution. At anode liberation of free electrons (oxidation) takes place resulting in metal corrosion and reduction of the metal thickness. At cathode there is gain of electrons (reduction) and here the dissolved constituents in the conducting liquid accept the electrons to form nonmetallic ions. Metallic ions at anode and nonmetallic ions formed at cathode move through the conducting medium forming corrosion products.

3.5.3   Methods for Corrosion Control

3.5.3.1   Cathodic Protection (CP)

Cathodic protection is one of the most effective methods for preventing galvanic type of corrosion which can occur when two different metals are put together and exposed to a corrosive electrolyte. Without CP, metals act as the anode and easily lose their electrons and thus they get oxidized and corroded. In this method, metal surface to be protected is made to behave like cathode by providing electrons from another source. This is based on the principle that in electrochemical corrosion, anode is the one which undergoes corrosion and not the cathode. In other words, if we want to protect a particular metallic structure, we can create conditions where this metal becomes the cathode of an electrochemical cell.

The oil and gas industry, in particular, uses cathodic protection systems to prevent corrosion of fuel pipelines, steel, storage tanks, offshore platforms, and oil well casings. Another common application is in galvanized steel, in which a sacrificial coating of zinc on steel parts protects them from rust.

Cathodic protection can be done by two ways:

Sacrificial anodic connection: In this method metal surface of equipment to be protected is connected by a suitably sized wire to a small block or piece of more anodic metal such as magnesium, zinc, aluminium and their alloys (*which are more reactive*). The latter, usually called the galvanic or sacrificial anode, has a less negative electrochemical potential compared to the metal component being protected. Therefore, the sacrificial anode undergoes oxidation rather than the operating equipment and gets corroded and thereby protects the parent (cathodic) structure. This sacrificial metal corrodes preferentially (acting as the anode) while the more valuable metal object (structure to be protected) under consideration (acting as the cathode) remains protected. Main advantage of using this method is that no external power is required and is easy to install (Fig. 3.1).

In brief, galvanic cathodic protection involves protecting a metal surface of a piece of equipment by using another metal that is more reactive. Normally metals such as aluminium, magnesium, and zinc are used as sacrificial anode in industrial practice.

Impressed current cathodic protection (ICCP): In this method DC current is applied with negative terminal connected to metal to be protected and positive terminal connected to the anode. The steel component is thus connected to the negative terminal of the power source and the impressed current anodes are connected to the positive terminal of the power source. By doing so, the metal to be protected gets converted from anode to cathode thereby protecting it from corrosion. In ICCP, electrons are supplied to the cathodic structure using an external DC power source. Anodes such as graphite, high silica iron, stainless steel, and platinum can be used.

ICCP is a more economical method of CP when underground pipelines are long or offshore equipment is too large to protect via one or few galvanic anodes. This can be used for protection of large structures, water tanks, buried water and oil pipelines, and condensers.

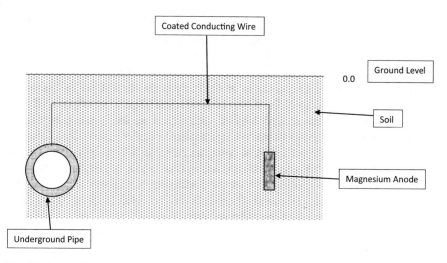

Fig. 3.1 Schematic arrangement for sacrificial anode

3.5.3.2 Anodic Protection

Anodic protection is a kind of corrosion protection designed to protect metals exposed in highly corrosive environments that are either too acidic or too basic for metals.

In anodic protection, the metal surface to be protected from corrosion is made more anodic by connecting to potentiostat and applying appropriate anodic currents to the metal. *This shall be assessed by lab trials.* The metal or alloy is protected due to growth of highly protective oxide surface film by application of anodic current on metal. Here a potentiostat is used to maintain the metal at constant potential with respect to reference electrode. The potentiostat shifts the corrosion potential of metal into passive potential so that corrosion is controlled. Anodic protection is used only for metals that exhibit passivity.

Anodic protection is commonly used in industries with underground facilities. This is used in chemical reactors, tanks and pipes carrying corrosive liquids and storage tanks for sulphuric acids, chromium in contact with hydrofluoric acid, and 50% caustic soda where cathodic protection is not suitable due to very high current requirements.

The main advantages of anodic protection method are that low currents are required and with this method large reduction in corrosion rate is possible.

3.5.3.3 Corrosion Inhibitors

Inhibitors are chemical compounds (organic or inorganic compounds) which are added in small quantities to a liquid or gas to decrease the corrosion rate of metal or the alloy which can come into contact with the process fluid. Inhibitors are usually dissolved in aqueous solutions. They reduce corrosion by either getting adsorbed as a layer on the entire surface of the corroding metal or forming a protective film which prevents corrosion by acting as a barrier. Organic inhibitors are in general adsorbed on the metal surface and can slow down the diffusion of H^+ ions.

Inhibitors can be either anodic or cathodic by controlling either the anodic or cathodic corrosion reactions. Examples of anodic inhibitors are chromate, silicate, and nitric salt of sodium and potassium. Arsenic and antimony ions are cathodic inhibitors and retard hydrogen evolution reactions in acid solutions. Zinc, magnesium, and calcium salts are the most common examples of cathodic film forming inhibitors. Sodium sulfite and hydrazine act as scavengers by removing corrosive reagents from aqueous solutions and are used as oxygen scavengers. Organic inhibitors such as amines, aldehydes, and alkaloids act as both anodic and cathodic inhibitors based on their reaction at the metal surface and potential.

Corrosion inhibitors are used to protect equipment in oil well extraction and processing, petroleum refineries, and chemical manufacturing.

Factors affecting choice of inhibitors are cost, toxicity, availability, etc. The inhibitors should not form any harmful deposits on the metal surface.

3.5.3.4 Protective Linings for Equipment Shells

Inner protective linings are provided for protection of equipment shell from the (reaction) materials which could be at high temperatures or have corrosive/erosive properties *because these materials can reduce the life of the equipment drastically.*

Protective linings are also provided to minimise the contamination of the process material due to reactions with the equipment shell.

Thermal linings also serve to reduce the heat losses during operation and thus improve the thermal efficiency of the process plant.

Fibreglass reinforced plastic, PTFE, polypropylene, and rubber are the commonly used lining materials for protection against dilute solution of acid/alkali, etc. The linings are cured as per instructions from manufacturers; and their integrity is checked by electrical spark test at 5000 V.

Samples of metallic pieces coated with appropriate thickness (1.5–5.0 mm) of the lining material shall also be tested by exposure to actual operating conditions.

However, these lining materials can get eroded if the reaction media in contact with them contain sharp, hard particles.

Use of Fibreglass reinforced plastic and rubber linings is to be avoided at temperatures more than 65 °C or in contact with concentrated acids as they can get damaged. PTFE (Teflon) lining is preferable.

Glass linings are also used for process vessels to prevent contamination during handling of special chemicals. They are not suitable for hot phosphoric acid and HF, should be very carefully handled, and shall be only very slowly heated or cooled by control of the temperature of the heating/cooling medium in the jackets.

A combination of lining materials can also be provided to achieve the desired results.

Refractory Lining

Refractory compositions have oxides such as SiO_2, Al_2O_3, CaO, MgO, ZnO, TiO_2, Cr_2O_3. Ceramics are also used as refractory materials as they are resistant to heat and corrosion.

Thermal protection materials (bricks) for shells of furnaces / process units operating at high temperatures are chosen on the basis of the following:

- Capacity to withstand maximum temperature
- Thermal conductivity
- Coefficient of thermal expansion
- Apparent porosity

Lower porosity indicates a dense brick with higher thermal conductivity, and higher cold crushing strength CCS

- Alumina Al_2O_3 content.... higher alumina content is desirable. Generally, it is 40–45% and above up to 65–70%. Special refractory bricks are also manufactured up to 90–95% alumina content
- Fe_2O_3 content shall be generally less than 2.0–2.5%

Refractory materials (bricks) with higher iron oxides content can get reduced in the presence of CO or SO_2 and may develop cracks in the lining. If SO_2 is present it can permeate through refractory lining and react with the steel shell to corrode it. Hence an additional layer of refractory lining is provided along with ceramic paper or by applying a layer of silicate cement. The ceramic paper also further reduces heat loss from the equipment.

Refractoriness Under Load (RUL): This is the capacity to withstand weight at high temperatures. It is to be considered for tall internal linings where the bottom layer is subjected to load from the top.

Acid-resistant Bricks Lining

- These are selected after determination of loss of weight of the sample after subjecting it to the acid to be handled for about 24–48 h at temperature up to 120 °C.
- The porosity shall not be more than 1.5%.
- Surfaces of the AR bricks shall be salt glazed to minimise penetration of the corrosive medium.
- Cold compressive strength (CCS) shall be sufficient to allow use as lining material in tall towers. Test certificate for CCS shall be obtained from manufacturer when the bricks are to be used as lining material in tall towers.
- Acid-resistant cements and mortars are to be used to join the bricks.
- It is advisable to use fire bricks and acid-resistant bricks with curved shape when they are being used for lining of vessels up to 2000 mm ID.
- These bricks with standard rectangular shape can be used for the lining of vessels with ID more than 2000 mm.
- Only properly shaped (curved) arch bricks shall be used for lining of manhole nozzles. This is necessary for stable brick lining at the manholes.
- In earlier practice the faces of adjacent fire bricks used to be made wet before installing so that they did not absorb moisture from the binding cement paste. This was to allow proper setting of the cemented joint. However, in recent practice this may not be necessary. It is better to consult the manufacturer of the cement.
- It is preferable to use only clean filtered water for making the cement paste.
- Insulating and fire bricks shall never be exposed to rain. They should be kept in a covered area.
- As per Indian Standard IS 4860 the cold compressive strength (CCS) of acid-resistant bricks should be at least 700 kg/cm^2 when used for lining of tall towers. Their porosity shall not exceed 1.5%. The resistance to acids as measured by loss of weight should not exceed 0.5%.
- It is preferable that they should have salt glazed smooth surfaces (Table 3.3).

3.5.3.5 Cladding

Claddings are widely used in chemical processes, oil refining, and electric power generation industries. This can save considerable cost of using only alloy. Carbon steel can be clad with a more corrosion-resistant alloy.

This can be accomplished in several ways such as roll bonding, weld overlaying, and wallpapering. Cladding on sheet metal is achieved by passing both metal and protective cover through rollers at high temperatures and both are bonded together. For forged components such as nozzles and flanges, cladding is done by weld deposition.

Cladding austenitic or ferritic stainless steel on base metal of carbon steel gives a good corrosion resistance and improved mechanical properties. Materials such as titanium or zirconium, glass, lead, and polymers such as PTFE and FRP are used for better corrosion resistance. A variety of plastic polymers can be used to provide protection and extended service in severe corrosive working conditions, offering excellent resistance as well as stability at medium range of temperatures.

Glass cladding provides good corrosion resistance to saline solutions, organic substances, halogens (except hot phosphoric acid and HF), and acidic and alkaline mixtures. It provides better protection and resistance to chemical attack even during prolonged periods of exposure and at temperatures generally below 100 °C and can prevent contamination of reaction materials. However, the glass-lined equipment should not be subjected to vibrations or rapid heating and cooling.

Limitation of using clad construction is the possibility of failure of the fillet welds which are used to make joints in the liner thereby releasing corrosive compounds throughout the backing material causing undetected corrosion. Also, clad construction is difficult and relatively expensive compared to solid construction for complicated structures with many nozzles, attachments or complex internals.

3.5.3.6 Hot-Dip Galvanization

This corrosion prevention method involves dipping steel into a bath of molten zinc. The steel gets coated with the zinc to create a tightly bonded alloy coating which serves to protect the steel. However, this cannot be done on-site. There is also an environmental pollution issue because the zinc fumes which release from the hot galvanizing bath are toxic.

3.5.3.7 Corrosion-Resistant Coatings

Thin coatings of metallic and nonmetallic materials provide protection from corrosion. Metal coating can be applied by electro-deposition, cladding, hot dipping, vapor deposition, etc. In electroplating method, coating of metal is deposited on metal surface by passing direct current through electrolytic solution. Hot dipping involves immersing metal to be coated in a bath of molten coating metal. Metal spraying is done by coating metal in molten state on the base metal. Inorganic coating can be done by chemical or electrochemical conversions at metal surface, spraying or diffusing.

Organic coatings such as paints, enamel, and varnishes are also used for the protection of base metals.

Phosphate coating is a coating applied on the ferrous metals to inhibit corrosion. In certain applications phosphate coating is followed by oil coating for improving its corrosion resistance. Zinc phosphate is widely used as a corrosion-resistant coating on metal surfaces and is applied as a primer. It has become a commonly used corrosion inhibitor and is more popular than the toxic materials based on lead. Zinc phosphate coats better on a crystalline structure than bare metal with use of a seeding agent for pre-treatment. However, a thick phosphate layer can cause splitting of paint. Hence the thickness of phosphate layer shall be kept low.

3.5.3.8 Corrosion-Resistant Paints

Corrosion-resistant paints serve as a coating to protect the metal surface from corrosive compounds. Modern paint systems use a combination of different paint layers. The primer coat acts as an inhibitor, the intermediate coat increases the overall thickness of the paint, and the final finishing coat serves as a resistance to environmental conditions.

Base coat is epoxy based such as (zinc rich epoxy primers, yellow zinc primer). It is then followed by two coats of corrosion-resistant paints like chlorinated rubber paints or bituminous paints. A combination of above is used as per the severity of conditions.

However, the coats of paints need regular maintenance as they can peel off during service life. This can expose the metal to corrosive conditions. It is generally necessary to remove all old coats of paint and apply fresh coats again.

It is advisable to consult the manufacturer of the coatings and paints for selection of the correct protective material. A small test piece of the metal to be protected

Table 3.3 Examples of refractory materials

Refractory material	Uses
Silica bricks	In roofs of open-hearth furnaces, glass furnace, coke oven walls, linings of acid converters
Fire clay bricks	Being cheaper are used for construction of blast furnaces, open hearth furnaces, stove ovens, crucible furnaces, kilns in zones operating below 1000 °C
High alumina bricks	Cement rotary kilns, lower parts of soaking pits, brass melting reverberatories, lead dressing reverberatories, aluminium melting furnaces, combustion zones of oil fired furnaces
Magnesite bricks	Used in steel industry for linings of basic convertors, open hearth furnaces, reverberatory furnaces for smelting lead, copper and antimony ores, hot zones of cement rotary boilers, refining furnaces for gold, silver, and platinum

shall be coated with the corrosion-resistant coating and paint; and then subjected to the actual working conditions. The metal surface shall be observed after 120 h to see the protective effect.

3.6 External Insulation Material

External insulation is provided to minimize heat loss from the surface of the shell and also to protect the shell of the equipment from corrosive atmosphere which may be present in the plant. It also protects workers from burns if they accidentally come in contact with the shell.

As per Stefan–Boltzmann law, the heat loss by radiation is a function of fourth power of the absolute temperature of the surface. The loss will be small in case of low temperature difference between the surface of the (hot) pipes and the surroundings. However, as the temperature difference increases, the loss increases rapidly. Hence it is important to minimise the temperature of the external surface and also to keep it shining.

Air is a very poor conductor and heat loss by conduction will therefore be small except through the supporting structures. However, convection currents form easily and the heat lost from the unlagged surface is considerable.

The conservation of heat is therefore an economic necessity by providing insulation and some form of lagging on the hot surfaces.

Thus the main purpose of thermal insulation on *external* surfaces is:

- Conserve heat energy of inside process material.
- To enable efficient operation of downstream process units (for better process control).
- To control temperature of exposed surface (to protect the employees from burns).

The temperature of the metallic shell of the process units and large ducts operating at high temperatures can be reduced by providing *internal* lining by refractory and insulating bricks; by castable refractory layer or by providing ceramic paper. However, composition of the hot gases and the dew point (in case of presence of moisture and acidic gases) should be considered to avoid condensation of acidic droplets on the metallic shell from inside. Hence in certain cases the internal insulation is designed to limit the shell temperature above the dew point; and external insulation is provided thereafter to minimise the heat loss.

These aims can be achieved by having appropriate thermal resistance to heat flow. Thermal resistance is that property of insulating material that opposes passage of heat through the material.

3.6.1   Design of the Thermal Insulation

This shall be carried out while considering the below:

1. Thermal resistance is directly proportional to thickness of insulating material and inversely proportional to thermal conductivity of the insulating material.
2. Heat flow directly depends on the temperature difference between hot surface and surroundings; and inversely proportional to thermal resistance.

The selection of the best insulating material for a particular service condition will depend on the efficiency of various types and forms of material in that particular temperature range.

The type of thermal insulation to be used can be classified according to temperature range as:

Low temperature: This is used for temperatures below room temperature and particularly used in refrigeration service.

Medium and higher temperatures: Skin temperatures of process vessels/hot gas pipelines are not allowed to go to 450 °C by providing internal lining of refractory/insulating bricks or ceramic paper.

3.6.2   Factors for the Selection of Insulation Material

• Ability of the material to withstand the working temperature range.
• Should have low thermal conductivity.
• Should have a type and form which has sufficient durability and structural strength to withstand several conditions such as moisture and chemical environment.
• Ease of installation/application to ensure the retention of original form.
• Cost of material for selecting the most economical insulation.
• Consider the fixed cost against the heat saved

3.6.3   Various Insulation Materials

• Glass wool (or fibre glass): This non-combustible material is made from selected glass in the form of fine fibres which are light in weight and generally used in every type of industry to conserve heat energy. The material is used in the loose form or in blanket form. It can be used for temperature range of −150 °C to +500 °C.
• Mineral wool: It is manufactured by blowing molten silica into threads. These threads can be converted into insulated blankets by adding suitable binding material. When a waterproofing material is used, an effective low temperature

insulation can be produced which is highly moisture resistant and can also be used for insulation of buildings. Mineral wool can be used up to +600 °C.

- Slag wool: This can be used up to +600 °C.
- 85% Magnesia: It consists of hydrated magnesium carbonate with asbestos fibres and binders. It is generally available in rigid blocks but can be obtained in plastic form also. It can be used up to a temperature of 315 °C for valves and pipe fittings.
- Calcium silicate: It is also available in both rigid and plastic form like 85% magnesia for use in pipeline and pipe fittings up to temp of +650 °C. It is also available in light weight blocks.
- Asbestos: Asbestos has varying amounts of MgO, SiO_2, FeO, etc. *However, use of this material is banned in many countries.*
- Diatomaceous earth material: Unusual form of silica with a high melting point, 1600 °C. Used as pipe insulation for temperature ranges from −10 °C to 1000 °C.
- Polystyrene: This is a lightweight synthetic material. This is suitable for sub-zero temperature insulation. This has very low thermal conductivity.
- Cork: This is a combustible material and hence is used only for low temperature service lines. Temperature range −160 °C to 90 °C. *However, the use of this material may not be allowed in process plants which handle inflammable materials.*

External Cladding

The external insulation layer is covered by aluminium sheets (generally SWG 20-22) to prevent ingress of moisture during rains. They minimise radiation losses as they are shining.

Care should be taken to provide perfect joints in these sheets. It is advisable to regularly monitor the temperature of the external surfaces either by direct measurement or by thermography imaging to detect inefficient insulation.

References/for Further Reading

1. "Corrosion Engineering", Mars G. Fontana, McGraw-Hill Book Company, Third Edition.
2. "Corrosion Engineering – Principles and Practice", Pierre R. Roberge, McGraw-Hill,2008.
3. "Handbook of corrosion Engineering", Pierre R. Roberge, McGraw-Hill.
4. "Theory and design of pressure vessels", John F. Harvey, P.E. CBS Publishers
5. "Joshi's Process Equipment Design for Chemical Engineering", V.V. Mahajani and S.B. Umarji, Laxmi Publications.
6. "Process Equipment Procurement in the Chemical and Related Industries", Kiran Golwalkar, Published by Springer International Publishing, Switzerland.
7. "Production Management of Chemical Industries" Kiran Golwalkar, Published by Springer International Publishing, Switzerland.

Chapter 4
Pressure Vessels

Abstract Pressure vessels are used in many chemical industries for carrying out reactions, storage of gases and liquids, for heat recovery, etc. They are very important for operation of plants. It is necessary to specify the function, operating conditions, material of construction, etc. while designing and fabricating. Fabrication codes and standards should be strictly followed to avoid any failure. Economic considerations are also very important for selection of material of construction for pressure vessels.

Adequate safety devices shall be provided on pressure vessels.

Post weld heat treatments (PWHT) are to be given to relieve the stresses introduced during welding done while fabricating. The vessels are to be tested as per standard procedures for radiography, hydraulic tests, etc. Important documentation includes design, fabrication drawings, test results of materials used, details of PWHT, pressure tests carried out, etc. Procedures recommended by manufacturer should be followed for installation and commissioning. Proper isolation and flushing out should be done prior to taking up any maintenance job. The vessel shall be pressure tested again before taking in to use after repairs.

Keywords Classification of pressure vessels · Considerations for design and selection of pressure vessels · Codes and standards · ASME boiler and pressure vessel code · Maximum allowable operating and working pressure · Types of heads · Material of construction for pressure vessels · Temperature limitations · Protective devices · Pressure relief and safety valves · Inspection guidelines · Visual test · Radiographic test · Hydrostatic and pneumatic tests · Post weld heat treatment · Important documents · Pre-commissioning checks · Operation of pressure vessels · Maintenance procedure · Additional precautions

© The Author(s), under exclusive license to Springer Nature Switzerland AG 2022   55
K. R. Golwalkar, R. Kumar, *Practical Guidelines for the Chemical Industry*,
https://doi.org/10.1007/978-3-030-96581-5_4

4.1 Introduction

Materials can be processed or stored in vessels such as cylinders, tanks, towers, reactors, drums, dryers, hoppers, bins, and other similar containers. Size and shape of vessels vary depending on the requirements of the process. Tanks are used to store products in large quantity while reactors are used for carrying out chemical reactions required as per the processes used. Towers, often the tallest vessels in a plant, are used where separation or stripping processes take place. Many of these units are to be operated under considerable pressure in batch or continuous mode. These vessels are to be very carefully designed, fabricated, and operated to ensure safe working in the chemical plants.

Examples of pressure vessels used in chemical industries are storage of compressed chlorine and hydrogen in cylinders; vessels for dissolved air floatation units in effluent treatment plant (ETP), receivers for supply of compressed air; waste heat recovery systems (for generation of steam); steam jacketed pressure leaf filters, reactors, and condensers in liquid SO_2 production plants; and liquid SO_3 condensers and storages. Typical pressure vessels in the oil/gas industries include vessels for separation of crude oil and gas at oil wells, heater treaters and desalters for treating crude oil, and vessels for separation of oil from produced water (water which comes along with the crude oil).

4.2 Classification of Pressure Vessels

Pressure vessels can be classified as:

- Based on type of pressure the vessels are subjected to:

  - internal pressure (storage vessel, heat exchanger)
  - external pressure (jacketed vessel)

- According to geometrical shape (cylindrical, spherical, conical, ellipsoidal)
- Whether an open vessel (surge tanks) or a closed vessel (heat exchanger)
- According to dimensional ratios (i.e. ratio of wall thickness to vessel diameter)

  - Thin walled (dimensional ratio <1: 10): distillation columns, heat exchangers
  - Thick walled (dimensional ratio >1: 10): autoclaves, nuclear reactors

- Based on type of heating system of the vessels:

  - Fired pressure vessels: These are closed containers that are directly in contact with some heating source.
  - Unfired pressure vessels: These are closed containers that are not directly in contact with some heating source but are subjected to internal pressure and used for storing, receiving, or carrying fluids under pressure. Examples are condensers (with incoming vapours at high pressure), compressed air

receivers, inert gas cylinders, vessels carrying dissolved acetylene, and those which may carry LPG.

- Other pressure vessels in chemical plants: These are vessels which can get pressurised due to (1) failure or insufficient supply of cooling medium, (2) runaway reactions due to excess feeding of reactants, (3) exothermic reaction which may occur due to ingress of moisture which may pressurise the system, (4) storage tanks which can get pressurised due to high ambient temperatures in summer, and (5) leaking of water-cooled electrodes or jackets of process vessels operating at high temperature.

4.3 Important Considerations for Design and Selection of Pressure Vessels

Design calculations and specifications for all major equipment are made by the designer. Wherever possible, standard equipment (if available) may be selected as they can be procured at reasonable low cost as well as with known performance if they are being used at some other plants. But if specially designed equipment are needed, a detailed specification sheet must be presented to the fabricator /supplier.

These sheets should specify the following:

- Intended use (function of the vessel): As reactor, condenser, heat exchanger, boiler, air receiver.
- General arrangement (GA): Drawing giving main dimensions, type of heads, nozzles (*see below*), corrosion allowance, mountings and other arrangements, supports required.
- Duty cycles and conditions: whether it is to be used intermittently or continuously.
- Normal operating pressures (internal/external).
- Test pressure
- Maximum operating pressure likely during course of operation.
- Chances of any sudden rise of pressure and/or temperatures during operation (due to runaway reaction, failure of cooling water flow, increase in heating by malfunctioning of temperature controller) should be considered at design stage itself.
- Maximum and minimum operating temperatures (in order to select the material to be used for construction).
- Requirement of radiography.
- Frequency of pressurisation/depressurisation during use of the vessel.
- Physical and chemical properties of materials to be handled by the vessel.
- Composition of materials to be handled and presence of suspended solids (since these can choke the connecting nozzles for safety valves, pressure taps).
- Appropriate material of construction (safety and economic considerations).

- Installation position—on ground floor/at a height.
- Location in plant at which it will be installed i.e. whether (indoor/outdoor).
- Nozzles and fittings required for incoming and outgoing process materials; safety vent, safety valves, pressure and temperature indicators, drain valve, inspection and cleaning manholes, entry of agitator shaft, observation nozzles, level indicator, sight and light glasses, sampling points, for draining out the vessel.
- Orientations of nozzles.
- Details of arrangement for heating / cooling: Whether by external heating/cooling jacket or limpet coils; heating medium which will be used and maximum temperature of the heating medium.
- Volumetric capacity required for the vessel (They are generally not filled up more than 85% when LPG is stored to keep space in the vessel for vapour/expansion of the liquid).
- Insulation required.
- Externally connected pipes and ducts (These should have independent supports to prevent mechanical stresses on the nozzles).
- Corrosion from inside and outside due to corrosive and reactive properties of the contained fluid and the surrounding atmosphere.
- Erosion caused by high velocity of fluid flow (inside: due to agitation, outside: due to fluid in jacket).
- External loading due to wind, snow, and piping system attached to the pressure vessel.
- Provision of lifting lugs, support legs, and cleats for fixing external thermal insulation.
- Shaft seals for vessels with agitators: Select a suitable mechanical shaft seal for agitators as per operating conditions, and properties of material inside the vessel (inflammable, explosive or toxic nature). Some other options are: conventional glands and water-cooled glands (for low pressure only).
- Allowable tolerances in certain dimensions or nozzle orientations. *Any reasonable deviations suggested by the fabricator can be discussed with the designer.*
- Any specific conditions for quality assurance plan (QAP) which the fabricator must comply with (e.g. welding electrodes to be used).
- Number of pressure vessels required may also be mentioned. This may reduce the quoted price.
- Stress relieving requirement shall be mentioned.

4.4 Design of Pressure Vessels

Design of pressure vessels incorporates the calculation of dimensions of the unit and analysis of induced stress /strain apart from selection of proper material of construction.

4.4.1   *Pressure Vessel Codes and Standards*

Pressure vessels are employed to contain and store various liquids and gases which may be highly inflammable or toxic. Hence it is imperative to handle these chemicals with utmost care and follow safety protocols as mentioned in the codes and standards. The codes and standards govern the procedure for design, material of construction, fabrications, inspection, testing, operation of pressure vessel, and safety. The national standards and codes in the country of use must be followed. In their absence one should follow international codes and standards.

Manufacturers of pressure vessels must follow codes and standards in force in the country of manufacture. Pressure vessels manufactured for export must comply with codes and standard in the country of installation.

Some well-known codes and standards for pressure vessels are:

- Bureau of Indian Standards (BIS)
- ASME Boiler and Pressure Vessel code Section VIII: Rules for Construction of Pressure Vessels.
- American Society for Testing Material (ASTM)
- Pressure Equipment Directive (PED) by European Union (EU)
- Japanese Industrial Standard (JIS)
- German Pressure Vessel Codes (DIN)
- British Standard Specifications 5500

Indian standard for pressure vessels covers minimum construction requirements for the design, fabrication, inspection, testing, and certification of pressure vessels in ferrous as well as non-ferrous metal.

- Indian Boiler Code (IBR)—Design of steam boilers having capacity exceeding 25 liters.
- IS 13490—Code of practice for handling of speciality gases (flammable, corrosive, anesthetic, oxidising, etc.)
- IS 15578—Code of practice for handling of gas mixtures.
- IS 2825—Design of unfired pressure vessels.

Statutory Approvals in India

Any vessel or pipe that handles pressure more than 1.5 kg/cm^2 (g) is covered by statutory rules and all formal approvals are to be taken as given below.

- The drawings along with the calculations while designing the pressure vessel have to be approved, signed, and stamped by statutory authority before releasing for manufacturing. It is a provisional clearance to proceed with manufacturing.
- All other documents such as certificates for raw materials, procedures followed, and tests carried out are also to be approved as a proof that the manufacturing is in accordance with the laid out statutory rules and regulations.
- Final clearance document before dispatch and as-built drawings are also to be approved, signed, and stamped by statutory body inspector.

- The pressure vessel details have to be embossed on the vessel and impressions are to be taken. These are part of the final documents.
- All certificates, charts, rub-offs, as built drawings become part of the statutory record folder from manufacturer's location, which is then to be submitted for inspection and records to the local statutory authorities where the pressure vessel is to be installed for use.
- All these activities have to be completed before commissioning and authorisation for commissioning to be taken from local statutory body. If the statutory body has a reason to believe that the submitted documentation is not in line with the requirements or some clarification is required, then the manufacturer must provide the same to the satisfaction of the approving authority.

Note Statutory rules in force in the country where the pressure vessels are to be used must be complied with (Table 4.1).

As a matter of abundant precaution, PWHT shall be carried out on pressure vessels to be used for lethal materials, for operating at high temperatures or having thickness of more than 30 mms.

4.4.1.1   ASME Boiler and Pressure Vessel Code (BPVC)

The BPVC provides rules for the design, fabrication, installation, inspection, care, and use of boilers, pressure vessels, and nuclear components. The code also includes standards on materials, welding and brazing procedures and qualifications, and non-destructive examination. Some of the important sections are:

- **ASME BPVC Section I**—Rules for Construction of Power Boilers
- **ASME BPVC Section II**—Materials

  - Part A—QASR46 Ferrous Material Specifications
  - Part B—Nonferrous Material Specifications

Table 4.1 Classification of welded vessels according to IS 2825--1969 for Unfired Pressure Vessels

Class of vessels	Description/application
Class I	• Used for highly dangerous, lethal (inflammable/toxic) substances • Operating temperature could be below −20 °C • All butt joints shall be fully radiographed • There is no limit on thickness of vessel
Class II	• Medium duty pressure vessels • Non-toxic substances are treated in class II vessels • Welded joints are spot radiographed
Class III	• Light pressure vessels, not exceeding 3.5 kgf/cm^2 vapour pressure and 7.5 kgf/cm^2 hydrostatic pressure • Temperature range of 0 to 250 °C • Welded joints are generally not radiographically tested • Maximum vessel thickness can be 16 mm

- Part C—Specifications for Welding Rods, Electrodes and Filler Metals

- **ASME BPVC Section III**—Rules for Construction of Nuclear Facility Components
- **ASME BPVC Section IV**—Rules for Construction of Heating Boilers
- **ASME BPVC Section V**—Non-destructive Examination
- **ASME BPVC Section VI**—Rules for the Care and Operation of Heating Boilers
- **ASME BPVC Section VII**—Recommended Guidelines for the Care of Power Boilers
- **ASME BPVC Section VIII**—Rules for Construction of Pressure Vessels
- **ASME BPVC Section IX**—Qualifications for Welding, Brazing, and Fusing welders
- **ASME BPVC Section X**—Fibre reinforced Plastic Pressure Vessels
- **ASME BPVC Section XII**—Rules for the Construction and Service of Transport Tanks

ASME BPVC Section VIII Sections VIII is the regulatory code for the construction of pressure vessels.

ASME BPVC Section VIII consists of three divisions as follows:

ASME Section VIII Division 1

- This division covers the mandatory requirements, specific prohibitions, and non-mandatory guidance for materials, design, fabrication, inspection and testing, markings and reports, overpressure protection, and certification of pressure vessels having an internal or external pressure which exceeds 15 psi (100 kPa).
- Pressure vessel can be either fired or unfired. The pressure may be from external sources, or by the application of heating from an indirect or direct source, or any combination thereof.
- There are various Subsections, A, B, C, of this Division giving general requirements and fabrication of pressure vessels by different methods; and various materials of construction. *The reader may kindly refer to them for further information.*

ASME Section VIII Division 2

Alternative Rules

- This division covers the mandatory requirements, specific prohibitions, and non-mandatory guidance for materials, design, fabrication, inspection and testing, markings and reports, overpressure protection, and certification of pressure vessels having an internal or external pressure which exceeds 3000 psi but less than 10,000 psi.
- The pressure vessel can be either fired or unfired. The pressure may be from external sources, or by the application of heating from an indirect or direct source as a result of a process, or any combination of the two.
- The rules contained in this section can be used as an alternative to the minimum requirements specified in Division 1.

ASME Section VIII Division 3

Alternative Rules for Construction of High Pressure Vessels
- This division covers the mandatory requirements, specific prohibitions, and non-mandatory guidance for materials, design, fabrication, inspection and testing, markings and reports, overpressure protection, and certification of pressure vessels having an internal or external pressure which exceeds 10,000 psi (70,000 kPa).
- The pressure vessel can be either fired or unfired. The pressure may be from external sources, by the application of heating from an indirect or direct source, process reaction, or any combination thereof.

ASME BPVC Section IX Section IX is the regulatory code for welding, brazing, and fusing jobs operations.

There are three steps in qualifying welders and welding procedure specifications to Section IX of the ASME Boiler and Pressure Vessel Code (BPVC).

1. Preparation of welding procedure specification (WPS).
2. Procedure qualification record (PQR) is used to verify the WPS.
3. The performance of the welders is verified by welding performance qualification test coupons. The variables and tests used with the particular variable ranges qualified are recorded on a welder's performance qualification (WPQ) record.

An abbreviated summary of items covered in ASME Section IX is provided below:
Article I—Welding General Requirements QW-100
Article II—Welding Procedure Qualifications QW-200
Article III—Welding Performance Qualifications QW-300
Article III covers responsibility, type of tests, records, welder identification, positions, diameters, and expiration and renewal of qualifications. Welders and welding operators may be qualified by visual and mechanical tests, or by radiography of a test coupon, or by radiography of the initial production weld.

Article IV—Welding Data QW-40-Article IV covers welding variables that are used in the preparation. Some of the welding variables are listed below:

- Joints
- Base materials
- Filler materials
- Positions
- Preheat
- Post weld heat treatment

4.4.2 Significance of Internal Pressure, External Pressure, and Design Pressure

Internal pressure (P_i) and external pressure (P_o) of the vessel must be taken into consideration while designing a pressure vessel.

- Failure due to internal pressure can occur when the stresses developed in the vessel exceed the strength of the materials. Yield strength properties are used to determine the permissible internal design pressure to avoid failures.
- During failure due to external pressure, the vessel loses its capacity to withstand its shape and undergoes an irreversible deformation having lesser volume than before.
- Thus, values of P_i and P_o are to be carefully considered for design.
- **Design Pressure**: It is the pressure used in the design calculation for the purpose of determining the wall thickness of various parts of the vessel. Wall thickness is calculated initially and then corrosion allowance is added to arrive at the actual wall thickness (Table 4.2).

*Notes **Maximum Allowable Operating Pressure (MAOP)** is a pressure limit set usually by statutory authorities (government body) which applies to compressed gas pressure vessels, pipelines, and storage tanks. **Never exceed this pressure during operation as a matter of safety.***

- *Maximum Allowable Working Pressure (MAWP) is defined as the maximum pressure based on the design codes that the weakest component of a pressure vessel can handle.*
- *Generally standard wall thickness materials are used in fabricating pressure vessels, and hence are able to withstand pressures above their design pressure. The MAWP is the pressure stamped on the pressure equipment, and the pressure that must not be exceeded in operation.*
- *Design pressure is the pressure for which a pressurised item is designed and is higher than any expected operating pressures. Due to the availability of standard wall thickness materials, many components will have a MAWP higher than the required design pressure.*
- *MAOP is less than MAWP*
- *Test pressure is the pressure at which the vessel is to be actually tested. Typically, the vessel may be tested at 1.5 times the maximum allowable working pressure or as per instructions from statutory authority.*

Table 4.2 Design pressure for vessels under internal and external pressures

Type of pressure in vessel	Design pressure
Vessel under internal pressure (Pi)	Design pressure = maximum working gauge pressure + 5% extra
Vessel subjected to external pressure (P_0) – On the outside of the vessel atmospheric pressure and inside vacuum corresponds to P bar	Design pressure = (1-P)
– Inside atmospheric pressure and external pressure above atmospheric	Design pressure = maximum external gauge pressure + 5% extra
– Internal pressure below atmosphere and external pressure above atmosphere	Design pressure = $P_0 + 0.05\ P_0 + (1\text{-Pi})$ or $(1.05\ P_0 + 1)$

4.4.3 Significance of Design Temperature

Determination of appropriate design temperature is important to find the allowable stress value of the material of construction which is temperature dependent (Table 4.3).

4.4.4 Heads of Vessels

Cylindrical shells /vessels are closed by covers or heads of different shapes, which can be joined to the shell by welding or flanged joints. These heads can be flat or formed ends such as elliptical, hemispherical, torispherical, and conical.

Flat covers are simplest type of head used as covers for manholes, heat exchangers, horizontal cylindrical storage vessels under light duty service of pressure below 1.5 kg/cm². The main disadvantage of flat head is the high discontinuity stresses at the joint between shell and the flat cover plate.

Formed ends are preferred for vessels under higher pressures in order to have a smooth change in shape and lesser stress concentration.

Torispherical heads are preferred for a pressure range of 1–15 kg/cm²g. In this type of head, the dish or crown radius is equal to or less than the diameter of the head. The knuckle radius at the corner should be designed to minimise stresses.

Elliptical heads are preferred for pressures over 15 kg/cm²g.

Hemispherical heads are the strongest among different types of formed ends for a given thickness but fabrication cost is very high. These are used for high pressure vessels.

Conical heads are preferred for equipment like evaporators, spray driers, and settling tanks to facilitate easy removal of thick liquor and solids.

In formed ends, the radius with which dish is formed is called crown radius (Rc) and the corner radius is knuckle radius (R_k). The straight part of the head (S_f) is selected so as to keep corner part away from the joint of head with the shell and to avoid stress concentration in the corner. The crown radius, knuckle radius, straight height of the heads, etc. are to be designed as per operating conditions, properties of the materials handled, properties of material of construction, and safety factor considered (when dangerous materials are handled) as per the guidelines given in codes. The design engineer/consulting engineer shall design the head after considering the

Table 4.3 Design temperature for different types of heating of vessels

Type of heating	Design temperature
For unheated parts	Highest temperature of stored material
For body parts heated by steam, hot water	Highest temperature of body +10 °C
Vessel with direct internal or external heating – vessel shielded – vessel unshielded	Highest temperature of inside material +20 °C Highest temperature of inside material +50 °C

above matters. These shall be fabricated accordingly and welded to the pressure vessel, stress relieved and radiographed. The pressure vessel shall then be tested as per statutory rules before commissioning (Fig. 4.1).

4.5   Selection of Material of Construction

Factors to be considered for selection of material:

- Service conditions: Maximum operating pressure and maximum and minimum operating temperature.
- Corrosive environment: Materials to be handled and their corrosive properties.
- Loading condition: Continuous or intermittent (cyclic operation of plant).
- Selection of reliable fabrication methods with due considerations for fabrication problems (if any) when a particular material is selected.
- Commercial availability of materials
- Cost of the material.

Commonly Used Materials of Construction for Pressure Vessels
Different materials of construction (ferrous and non-metals) and their properties have been discussed in Chap. 3 on materials of construction.

- Steel: It is the most widely used ferrous material of construction for pressure vessels. Carbon steel is used at ordinary to moderately elevated temperatures and non-corrosive services in many industries due to its strength, good ductility, and ease of workability. It has few alloying elements such as manganese, silicon, phosphorus, and sulphur in low concentrations and 0.05% to 0.6% carbon.
- Low alloy steel: This type of steel has carbon content less than 0.3%; manganese 1.0%, silicon about 0.3% and other alloying elements (Cr, Mo and Ni) which are less than 10%.
- High alloy steel is used for severe duty services. It is extensively used for corrosion resistance and for increased hygienic standard in construction of the vessel for avoiding contamination of the fluid by dissolved corrosion products or by the material itself. High alloy steels for construction of pressure vessel are ferritic stainless steels and austenitic stainless steel; duplex steel and super duplex steel (due to higher chromium and molybdenum content).
- Ferritic stainless steels, 11% to 30% chromium content, have good resistance to stress corrosion cracking.
- Austenitic stainless steels, 18% to 25% chromium and 8–20% nickel, have excellent corrosion resistance except corrosion by chlorides, nonmagnetic and have good strength and ductility.
- Nonferrous metals used as construction materials for pressure vessels include aluminum, copper, nickel, lead, tantalum, titanium, and their alloys (Table 4.4).

The mechanical strength of materials can reduce at high temperatures. Hence the designer should check the *allowable temperature* also for the material of

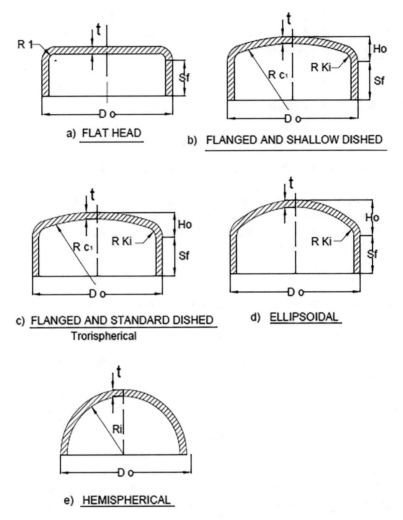

Fig. 4.1 Types of heads

construction while designing the pressure vessel (even though the limit for maximum temperature may be more for the particular material of construction). The vessel shall not be operated above this temperature (at the operating pressure).

Note   Approval shall be obtained for the design and fabrication drawing from statutory authorities.

4.6   Corrosion Allowance

Material loss by corrosion (i.e. reduction of thickness) is a very common problem in chemical industry. Additional thickness of metal known as corrosion allowance is added depending upon the operating conditions (to compensate for the loss).

Design for Reducing Corrosion

Proper design of equipment in process industries can help to reduce corrosion to a large extent.

Proper material of construction with good corrosion resistance properties should be selected. There are ways to inhibit corrosion such as having protective linings, cladding, and paints (discussed in Chap. 3 Materials of Construction).

Designer should avoid selecting materials which are likely to form galvanic couple and select material which are close in galvanic series.

Horizontal surfaces exposed to atmosphere have tendency to hold moisture and dust and should be avoided. Dished ends should be preferred to flat heads.

It is always advisable to have drains in the lowest point of a vessel to prevent accumulation of liquids. For large flat bottomed vessel proper sloping of bottom should be done for providing drainage. Sharp edges and crevices are more prone to accumulation of liquids and aggravate corrosion problems.

Welded joints should be preferred over riveted or bolted joints particularly for joining tube sheets in heat exchangers. Non-adsorbent gaskets such as Teflon are more resistant to corrosion. Wherever possible suspended solids should be removed by filtration, which would otherwise deposit on crevices and enhance corrosive conditions (Table 4.5).

4.7   Inspection Guidelines

All countries have their own codes and standards for pressure vessels which should be used for fabrication and inspection. Inspection procedures involve various tests depending upon the design of the pressure vessel and the nature of the liquid or gas contained. These tests are performed to detect punctures, cracks and stress developed during the fabrication process.

Table 4.4 Typical temperature limitations of various materials of construction

Material of construction	Temperature
Carbon steel	Non-corrosive service: Up to 400 °C
Stainless steels	Corrosive conditions: Up to 500 °C
Carbon molybdenum steel	Up to 550 °C
Chromium molybdenum steel	Up to 550 °C

4.7.1 Inspections during Visit to Fabrication Shop

- For stage inspection: Check the storage facilities for electrodes, plates, bought out components, thickness and composition of plates, cutting diagrams and cut plates, and weld groove preparations.
- The nozzle orientations for incoming/outgoing streams shall suit the layout of plant.
- Check assembly of vessel with only tack welding and root run of welding.
- Final visual inspection: Overall dimensions, pad plates and vent holes, lifting lugs, provisions for inlet and exit of process streams, safety vents/valves, and rupture disc; drain points, sample points, thickness of all flanges and the holes drilled on them, and gussets for nozzles.
- Final hydraulic test should be carried out as per statutory requirements. Use only calibrated standard pressure gauges and chloride-free treated water.

4.7.2 Visual Test (VT)

- Visual test can give good overview of the vessel.
- Pressure vessel inspectors must ensure that the vessel surface is clean and well-lit.
- Visual test shall be carried out only in good working conditions to observe for issues like cracking, corrosion, erosion, or hydrogen blistering.

  Limitations of Visual Test.

- Visual test allows only inspecting the surface of the vessel. Further inspection can be done by non-destructive testing methods only.
- Defects, if any, in the inner layers of the vessel cannot be determined.
- Generally, only large flaws can be found. Small minute cracks may not be noticed during visual inspection.
- There is a possibility of misinterpretation of the flaws.
- The accuracy of inspection depends on the experience and skills of the inspector as well as on their knowledge of welding.

Table 4.5 Corrosion allowance for different materials of construction

Type of alloy	Corrosion allowance
Carbon and low alloy steel	• 1.5 mm for non-severe condition • 3.0 mm for severe condition
Stainless steel and nonferrous alloy	No corrosion allowance
Wall thickness more than 30 mm	Corrosion allowance is not added

4.7.3 Radiographic Test

- Radiographic testing (RT) is a non-destructive examination (NDE) technique that involves the use of either X-rays or gamma rays to view the internal structure of a component.
- In the petrochemical industry, RT is often used to detect flaws in equipment such as pressure vessels and valves.
- RT is also used to inspect weld repairs.
- It is highly reproducible, can be used on a variety of materials, and the data can be stored for later analysis.
- Radiography is an effective tool that requires very little surface preparation.
- Moreover, many radiographic systems are portable, which allows for use in the field at elevated positions.
- RT should be carried out by trained and certified technicians only while using personal protective equipment to protect from radiation.

Types of Radiography
There are numerous types of RT techniques including conventional radiography and multiple forms of digital radiographic testing. Each works slightly differently and has its own advantages and disadvantages.

Conventional Radiography
- It uses a sensitive film which is exposed to the emitted radiation to capture an image of the part being tested.
- This image can then be examined for evidence of damage or flaws.
- The biggest limitation to this technique is that films can only be used once and they take a long time to process and interpret.

Digital Radiography
- Digital radiography does not require a film unlike conventional radiography. Instead, it uses a digital detector to display radiographic images on a computer screen almost instantaneously.
- It allows for a much shorter exposure time so that the images can be interpreted more quickly.
- Furthermore, the digital images are of much higher quality when compared to conventional radiographic images.
- With the ability to capture high quality images, the technology can be utilised to identify flaws in material, foreign objects in a system, examine weld repairs, and inspect for corrosion under insulation.
- The four commonly utilised digital radiography techniques in the oil and gas and chemical processing industries are computed radiography, direct radiography, real-time radiography, and computed tomography.

4.7.4 Pressure Tests

4.7.4.1 Hydrostatic Tests

- Hydro-testing is a process where components such as piping systems, gas cylinders, boilers, and pressure vessels are tested for strength and leaks.
- Hydro tests are often required after shutdowns and repairs in order to validate that the equipment will operate under desired conditions once returned to service.
- Hydrostatic testing is a type of pressure test that works by completely filling the equipment with water, removing the air contained within the unit, and pressurising the system up to 1.5 times the design pressure of the unit.
- The pressure is then held for a specific length of time to visually inspect the system for leaks. If the pressure reduces by itself it may indicate a leak.
- Visual inspection can be enhanced by applying either tracer or fluorescent dyes to the liquid to determine where cracks and leaks are originating.

Different Methods of Hydrostatic Tests
There are three common hydrostatic testing techniques that are used to test small pressure vessels and cylinders: water jacket method, direct expansion method, and proof testing method.

Water Jacket Method
- In order to conduct test by this method, the vessel is filled with water and loaded into a sealed chamber (called the test jacket) which is also filled with water.
- The vessel is then pressurised inside the test jacket for a specified amount of time.
- This causes the vessel to expand within the test jacket, which results in water being forced out into a calibrated glass tube which measures the total expansion.
- Once the total expansion is recorded, the vessel is depressurised and it then shrinks to its approximate original size.
- As the vessel deflates, water flows back into the test jacket.

The testing may result in permanent expansion of the vessel. This implies that the structure has lost its flexibility and is unsuitable for its intended function. The suitability of the vessel for the intended function then depends upon the permanent expansion of the vessel.

Direct Expansion Method
- Direct expansion method involves filling a vessel or cylinder with a specified amount of water, pressurizing the system, and measuring the amount of water that is expelled once the pressure is released.
- The permanent expansion and the total expansion values are determined by recording the amount of water forced into the vessel, the test pressure, and amount of water expelled from the vessel.

Proof Pressure Method
In this test an internal pressure is applied to determine if the vessel has any leaks or other weakness such as wall thinning that may result in failure during service.

4.7.4.2   Pneumatic Tests

- Pneumatic tests are performed on pressure vessels where presence of liquid is not acceptable.
- Pneumatic testing uses a non-flammable, non-toxic gas like air or nitrogen.
- A concern with pneumatic testing is that if a fracture occurs during testing for some reason, it could lead to an explosion.
- Pneumatic pressure testing of piping and vessels at moderate-to-high test pressures or at low test pressures with high volume is more hazardous than hydrostatic pressure testing because the stored energy is much greater in case of compressed gases than with water.
- A large amount of energy could get stored in compressed gas compared to water at the same pressure and volume. Hence when a joint, pipe, or any other component fails under test pressure when using compressed gas, this energy gets released with destructive force.

4.8   Stress Relieving by Post Weld Heat Treatment (PWHT)

- Stresses can develop in the pressure vessels during fabrication since cutting and welding operations are carried out. The welding process can therefore inadvertently weaken equipment by developing residual stresses (especially if cooled too fast after welding) which can lead to reduced strength. Hence the pressure vessels should be given a treatment known as post weld heat treatment to reduce the residual stresses.
- Different shapes of the welded parts (e.g. pressure vessel with many nozzles) can also introduce residual stresses at the welded joints. Such pressure vessels must be given PWHT.
- PWHT is necessary when the carbon and alloy content are more in the material used for fabrication of the pressure vessels, for vessels to be used for lethal substances and also for thick walled vessels.
- It is carried out in a special furnace (equipped with calibrated thermocouples) by heating the welded portion to a specified temperature for a given amount of time and then cooling at a controlled rate to reduce these stresses.
- Neglecting PWHT can lead to failure of welded joints. Hence it is a mandatory requirement for the fabrication and construction of pressure vessels.

Requirement for PWHT
There shall be proper facilities available with the vendor for PWHT as below:

- *A furnace with heating systems having calibrated thermocouples*
- *Automatic temperature recorder with proper control*
- *Proper loading and unloading facilities*
- *NDT facilities*

Precautions
- The welded portion to be given PWHT should be uniformly heated to the soaking temperature (as advised by the design engineer) and maintained for specified time.
- Thereafter it is to be cooled at a controlled rate in the furnace itself or in a slow current of inert gas.
- No cutting or welding shall be performed after PWHT.
- Inspection after PWHT must be done for welded joints or to check for any deformation of nozzles

4.9   Documentation

Important Documents
Copies of the following documents shall be maintained at plant and head office. This list is not exhaustive. Experienced engineers may add to the list if required.

- Design safety margin and corrosion allowance added
- Test certificates for all materials used for fabrication by approved test labs
- Traceability of material used (stampings on plates and original purchase bills for bought out items)
- Welding process followed
- Records of welding electrodes used
- Records of PWHT done for stress relieving
- Stage inspections and final inspection carried out.
- Records of hydraulic pressure tests carried out *using standard calibrated pressure gauges.*
- Each pressure vessel should have a nameplate fitted on it. It should give information such as serial number and date of manufacture, maximum allowable working pressure and temperature, and name of the manufacturer. A copy of document giving these details or a photograph of the name plate should also be preserved in the head office.

Statutory Documents to Be Preserved
These are required in many countries for obtaining permission to erect and commission the pressure vessels. They will be useful for future reference/records of the purchaser. Statutory authorities may recommend more documents to be preserved.

- Order copy giving intended use/service conditions for the pressure vessel
- General arrangement drawings and approved fabrication drawings with details of all nozzles and their orientations corresponding to plant layout
- PWHT records
- Radiography records
- Hydro-test records

- Test and guarantee certificates for bought components: safety valves, sight glass, level indicators, light glass, rupture discs
- Verifications of quality assurance plan done by inspecting engineers from purchaser.

4.10 Pre-commissioning Checks

- Carry out visual inspection of entire vessel before providing external insulation.
- Confirm orientation of all nozzles and cleaning manholes.
- Confirm the serial number, name plate, and last permissible date of service life.
- Use only properly calibrated pressure gauges.
- Check the support legs and structure for proper installation.
- Check installation of the protective devices and arrangements.
- Confirm setting of safety valves and rating of rupture discs.
- Confirm working of alternative (emergency) cooling arrangements if the vessel is to be cooled during operation.
- Installation must be on properly designed and constructed foundation.
- Vent holes at reinforcing pad plates must be kept clean.
- Provide audio visual warning systems.
- Do not use for different fluids than specified.
- Do not change colour code put on the vessel.
- Confirm smooth operation of valves (foreign objects may become lodged in valves during testing, despite efforts to remove them during cleaning).
- Where traces of water are not acceptable in process fluids, dry out the vessel, connecting lines and valves.

Note The pressure vessel can be commissioned after carrying out the pre-commissioning checks.

4.11 Operation of Pressure Vessels

- Study carefully the standard operating procedure before taking into use.
- Check calibration of pressure gauges before taking into use.
- Never exceed the maximum allowable operating pressure (MAOP)
- Set the safety valves as per statutory rules/less than MAOP pressure.
- Check setting of safety valves and safety interlocks regularly.
- Never exceed design operating temperature.
- Pressurise and de-pressurise as per OEM instructions only—never suddenly open/close valves.
- Drain out accumulated condensates (if any) periodically.
- Check thickness of vessel by ultrasonic instrument periodically.

- Generally, keep about 15–20% empty space when handling liquids which can generate considerable vapours.
- Carry out annual inspection and pressure tests on scheduled dates without delay.
- Do not use beyond end of service life.

4.12   Precautions

- Provide multiple safety devices—safety vents and valves, rupture discs, and high temperature and pressure cut off devices to disconnect power to fuel pumps (for preventing overheating).
- Protection must be provided from any spillages from nearby pipes carrying corrosive fluids
- Do's and don'ts recommended by original equipment manufacturer (OEM) and statutory authorities must be strictly complied with.
- Do not rapidly pressurise or depressurise.
- Never allow excess feeding of raw materials as it can lead to run away reactions
- There should be no interruption of cooling water/medium supply. Provide alternative arrangement also.
- Cooling jacket shall always remain full (even when supply of cooling water fails due to any reason). It should not get drained out by itself.
- Provide power by DG set to run the cooling water supply pump in case of power supply outage.
- Elevated storage reservoir should provide cooling water for temperature control in case of emergency.

4.13   Maintenance of Pressure Vessels

Pressure vessels are subjected to stresses and experience wear and tear during their service life. In order to ensure safety of the plant, routine inspection and maintenance of equipment is imperative. This holds especially true for pressure vessels, as their operating conditions are generally extreme. Maintenance should not be limited to detection and correction of faults. It must also include risk prevention.

Maintenance Procedure for Pressure Vessels
- Isolate the process streams and heating arrangement. Remove fuses from drive motors of process material feed pumps to prevent start up by mistake.
- Reduce the pressure slowly by venting out the pressurised contents through a scrubbing system. *Either collect carefully and reuse the material; or dispose through authorised party.*
- Now flush out any remaining vapours or gases by compressed dry air or nitrogen, whichever is suitable. The exiting gases are to be vented out through scrubbers.

- Check the gases being vented out by hand held instruments or suitable chemical detectors to confirm that all dangerous material has been removed before opening the vessel for inspection and repairs.
- Provide appropriate personal safety devices to the technicians. Confirm that suitable arrangements for firefighting, personal breathing apparatus, and rescue systems are in place.
- Obtain work permit from authorised senior plant engineer and get it checked by safety officer. Fresh permit is to be obtained every day before starting maintenance work.
- The inspection/cleaning manholes are to be carefully opened and the vessel is to be thoroughly cleaned from inside after removing any residual material in crevices.
- Check the vessel and all nozzles for any cracks, leaks, and weak gussets visually first. Now check it by hydraulic pressure at 1.5–2.0 times the working pressure for about 4–8 h (or as per instructions from statutory authority). Examine thoroughly if any drop in pressure is noticed.
- Check thickness of vessel walls by ultrasonic tester also.
- Clean the nozzles connecting the pressure vessels to safety valves, and for pressure measurement taps. Remove all dust, scales from inside thoroughly to remove any choke.
- Any weak spot, minor leak, weak gussets, or leaking portion is to be repaired by cleaning and welding.
- PWHT is to be carried out after repairs to *relieve stresses introduced during repairs*. The repaired vessel is to be offered for pressure testing to competent statutory authority again; and used only as per permission granted. Radiography of welded joints is a must when lethal fluid is handled or when the vessel will be used below minus 10 °C (-10 ° C) as per Indian standards

4.14 Special Equipment

4.14.1 Spherical Storages

Horton spheres are pressure vessels of a spherical shape.

They are used for storing LPG (liquefied petroleum gases) and other gases like propylene, nitrogen, and oxygen under pressure. The pressure is uniformly distributed on the shell. Hence the stress concentration is uniform all over the surface of the vessel.

Since the surface area is smaller for the same volume as compared to cylindrical shapes, the heat which can enter from the outside warmer surrounding is also less as compared to cylindrical storages. Hence the rise in internal pressure is also less.

Refrigerated storages are also used for storing gases at very low temperatures.

Design of these vessels depend on the pressure and temperature at which the gas is to be stored; the properties of the gas, the quantity to be stored, and mechanical strength of the material of construction should also be taken into consideration.

4.14.2 Gas Cylinders

Some typical rules generally applicable to cylinders handling gases under pressure are given below. However, this list is not exhaustive. The plant managements should obtain detailed list of rules prescribed by local statutory authorities in the country where the gas cylinder is being used. These must be complied with.

Marking on Gas Cylinders
- Identification marking by the manufacturer and owner.
- Marking by the local inspecting authority.
- Design specification to which the cylinder has been made.
- Mark of the heat treatment given to the cylinder e.g. PWHT
- Details of the hydrostatic test with the identification mark of test house.
- The date of test (month and year).
- Working pressure and test pressure.
- For LPG cylinders: Tare weight of the cylinder (which shall include the weight of the valve fitted to the cylinder) and water capacity.

Markings on Valves Fitted to the Cylinders
- The name or chemical symbol of the gas for which the valve is to be used
- Month and year of manufacture
- Marking by the local inspecting authority
- Identification symbol of the manufacturer
- Working pressure
- The type of screw threads on the outlet in case of left handed as (L.H.);

Precautions
Some important precautions to be taken while handling gas cylinders are:

- Always keep the cylinders upright.
- See that the valves are not hit by any object during handling.
- The cylinders should be securely placed on the trolley during shifting.
- Never drag them along the ground or pull by tying ropes on the valves.
- Unload a filled cylinder very carefully from the delivery vehicle. Do not carelessly throw it down.
- Keep the cylinders in a shed protected from direct sunlight.
- Do not change the colour of the cylinder.
- Do not use the cylinders after the last service date marked on it.
- Use only certified tested pressure regulators of approved manufacturers.
- The connecting hose/pipes should be pressure tested before taking into use. They shall not be used after permitted service life.

General

One must inspect all pressure vessels as instructed by statutory authorities. These vessels should never be continued in use beyond the last permissible date unless permitted in writing by statutory inspectors. During the maintenance cycle, all safety features on the equipment should be inspected and tested to make sure they are in proper functioning order.

Follow instructions from original equipment manufacturer (OEM) and statutory authorities. Test after the maintenance jobs are over and get it certified for use before connecting to process streams.

However, the vessel may be derated by the inspecting statutory authority after inspection. They may permit use only at a lesser pressure. A new vessel shall be procured in that case and the old one may be discarded or used for storing water, dilute alkali, etc. at atmospheric pressure.

4.15   Protective Devices

Installing protective devices on the pressure vessels ensures that the pressure vessel can be operated safely during working of the plant.

4.15.1   Pressure Relief Valves and Safety Devices

Pressure relief valve: It is a pressure relief device actuated by inlet static pressure and is designed to close again to prevent the further flow of fluid after normal conditions have been restored. Pressure relief valve is a generic term applying to relief valves, safety valves, safety relief valves, or pilot operated pressure relief valves, etc.

- Non-reclosing pressure relief device: A pressure relief device designed *to remain open after operation*
- Safety valve: A safety valve is a pressure relief valve characterised by rapid opening or pop action. It is used for gas or vapour service.
- Relief valve: A relief valve is a pressure relief valve, which opens *in proportion* to the increase in pressure over the operating pressure. Relief valves are generally used for liquids. In this type of valve, at the set pressure, the disk rises slightly from the seat without popping and permits a small amount of fluid to pass. As the pressure increases, the disk is further raised; thus an additional area is available so as to allow an increased flow of fluid.
- Safety relief valve: A safety relief valve is a pressure relief valve that can be used in either vapour or liquid service. For vapour service, it is adjusted to give a "pop" action, for liquid service it is adjusted for gradual opening.
- Pilot operated pressure relief valve: This is a pressure relief valve in which the major relieving device is combined with and is controlled by a self-actuated auxiliary pilot relief valve. The use of pilot-operated pressure relief valves may be

limited by the fluid characteristics (fouling, viscosity, presence of solids, corrosiveness) or by the operating temperature. The manufacturer should be consulted while buying it.

- Power-actuated pressure relieving valve: Movements to open or close are fully controlled by an external source of power (electricity, compressed air, steam, or hydraulic). If the power-actuated pressure-relieving valve is also positioned in response to other control signals, the control impulse to prevent over-pressure shall be responsive only to pressure and shall override any other control function. It has to be noted that the power-actuated pressure relieving valve *cannot be considered as safety device*, since, unlike the others, it relies on an external source of power.
- Rupture disc safety device: It is a non-reclosing differential pressure relief device actuated by inlet static pressure and designed to function by the bursting of a pressure-containing disk. A rupture disc device includes the rupture disc or sensitive element and the rupture disc holder. Rupture disc devices are used either alone or in conjunction with a pressure relief valve. The application of rupture discs alone is limited by the fact that when the disc ruptures the *entire contents of the system may be lost*. They may, however, be installed in parallel with a pressure relief valve to provide an additional safety. In this case, the relief valve is set at a lower pressure to limit rupture disc bursting during major pressure rise only.
- Rupture discs are pressure differential devices and the relieving capacity is therefore *affected* by the sizes and lengths of the inlet and outlet pipe lines.

It is advisable to connect the exit gas piping to a scrubbing system (with minimum pressure drop) to prevent environmental pollution when the rupture disc bursts open.

4.15.2   Other Pressure Release Systems

- Their design and construction shall be approved by statutory authorities.
- These shall be provided as close to the pressure vessel as possible.
- They should have locking arrangement of setting for release pressure i.e. should be tamperproof.
- They should function automatically.

Some of these are given below:

Breaking Pin Devices and Spring-Loaded Non-reclosing Pressure Relief Devices

A breaking pin device is a non-reclosing pressure relief device actuated by inlet static pressure and designed to function by the breakage of a load-carrying section of a pin which supports a pressure-containing member. A breaking pin device includes the breaking pin or load-carrying element and the breaking pin housing. Breaking pin devices shall not be used as single devices but only in combination with the pressure relief valve and the vessel.

Explosion Hatch It consists of a hinged metal cover placed over an opening in a vessel. The hatch consists of a hinged metal cover placed over an opening. It is used for vessels operating near atmospheric pressure and when the risk of explosion exists. Explosion hatches are not recommended for use at higher pressures, since the weight of the hatch will be excessive and this may prevent quick opening.

Liquid Seal Liquid seals can be used instead of pressure relief valves for set pressures below 10–15 psig, where relief valves are not considered reliable. Typical examples are the seal leg of a flare and liquid seals used as safety seals at inlet of condensers. The "U-tube" may be filled with water, mercury, or other liquid. Freezing of the sealing liquid shall be avoided by steam tracing or heating. Provisions for make-up and draining of the filling liquid should be made.

4.15.3 Safety Valves Provided by Vendor of Pressure Vessel

- Purchaser shall confirm the values for MAOP and test pressures from design engineer and statutory authorities while procuring the safety valves.
- Vendor should furnish test certificates with identification markings on the safety valves.
- The safety valves shall provide tight shut-off during normal operations (to prevent loss of process material inside the vessel through a leaking safety valve)
- At least two safety valves are to be provided as a matter of abundant precaution. They should be provided as close to the pressure vessel as possible.
- Reinforce the nozzles for mounting the safety valves in consideration of bending moment during blow off. These nozzles should be always kept clean (should have no deposit of solids inside)
- *Setting of safety valves shall be done as directed by statutory authorities.*
- Provide locking arrangement of setting for release pressure i.e. should be tamperproof.
- They should generally function automatically.
- Provide good lighting and easy approach for vent valves and pressure release valves so that they can be opened manually *if they do not open automatically during dangerous situations.*
- Should be of weatherproof construction and easy to maintain.

4.15.4 Additional Precautions

- Never use slings around nozzles while lifting. Use only lifting lugs
- Confirm that conditions prescribed in quality assurance plans have been complied with by the vendor.

- Close all openings by blinds during transportation from vendor's fabrication/workshop to prevent ingress of dust, moisture, or foreign bodies.
- Electrical tripping arrangements may be provided to trip the feeding pumps/compressors (for process streams or reactors) in case the pressure in the vessel becomes too high; in case of inadequate cooling of the inside materials due to any reason or runaway reactions in the vessel.

References/for Further Reading

1. "Integrated Maintenance and Energy Management for the chemical industries", Kiran Golwalkar, Springer International Publishers, USA.
2. "Joshi's Process Equipment Design for Chemical Engineering", V.V. Mahajani and S.B. Umarji, Laxmi Publications.
3. "Process Equipment Design and Drawing" Kiran Hari Ghadyalji, by Nandu Publications, Mumbai India.
4. "Pressure Vessel Design Manual", Dennis R. Moss, Gulf Publishing Company.
5. "Pressure Vessel Design Handbook", Henry H. Bednar, P.E., Second Edition, CBS Publishers & Distributors
6. "Process Equipment Procurement in the Chemical and Related Industries" Kiran Golwalkar, Published by Springer International Publishing, Switzerland.
7. "Production Management of Chemical Industries" by Kiran Golwalkar, Published by Springer International Publishing, Switzerland.

Chapter 5
Piping Design and Pumping Systems

Abstract Pressure drop in the piping can be reduced by increasing the pipe diameter. It enables to reduce the power consumption. This increase in cost of piping is to be balanced against reduced cost of power required for pumping.

The power consumption for pumping can also be reduced by proper design of piping layout (which should be convenient to operate and maintain) and by controlling speed of the drive motor by providing variable frequency drive. Colour coding should be used in a piping network to identify transfer of different fluids in a chemical plant to avoid confusion.

Various types of valves used for isolation and controlling fluid flow and the different types of actuators for operating them are also described.

The total pressure loss in piping systems is calculated by the sum of hydrostatic pressure difference, friction pressure loss, and losses due to sudden expansion, contraction, and minor losses due to fittings such as valves, bends, tees, and elbows.

It is necessary to properly install the pumps whether outside or submerged inside process tanks. Various types of pumps for handling fluids with corrosive properties, viscous nature, and carrying suspended solids are described in this chapter.

Problems during operation of centrifugal and positive displacement pumps have also been addressed.

Installation and operation procedure of centrifugal and positive displacement blowers are given.

Keywords Pressure drop in piping · Pipe schedules · Pipes of different materials · Colour codes for pipes carrying different fluids · Types of valves and actuators · Various types of pumps · Installation of vertical submerged and horizontal centrifugal pumps · Problems in operation of centrifugal and positive displacement pumps · Centrifugal and positive displacement type of blowers

5.1 Introduction

Piping is one of the important parts of a chemical plant. It is the medium used to transport process fluids and utilities from one unit to another. The piping system of a chemical plant shall be very safe to operate, economical to fabricate and install; and convenient to maintain.

For a given flow rate of fluid if a large size of pipe is selected then it can result in a lesser pressure drop. Larger size of pipe increases the fixed cost of pipe but the lesser pressure drop means lower power consumption. Lesser pressure drop in pipe may reduce the size of flow moving device like pump and thereby it may reduce the fixed cost of pump. Thus capital cost of pipe increases with diameter, whereas pumping or compression cost decreases with increasing diameter. Cost of piping is directly proportional to the diameter, length, and thickness of the pipe (schedule number). Energy consumption for pumping a fluid is proportional to the pressure drop, flow rate, length of the piping, density of the fluid, and friction factor. Cost of pumping is directly proportional to energy consumption.

Hence one should balance the extra cost of pipes (by selecting a higher diameter for the pipe) against the reduction in the of the cost of pumping (energy) and make the final choice by checking the payback period for recovering the higher cost which shall not be normally more than two years.

5.2 Factors to Be Considered for Piping Design

Process piping in a chemical plant is to be very carefully designed and fabricated. The properties of the fluids to be handled, maximum operating temperature and pressure; presence of suspended solids, permissible pressure drops, properties of the materials of construction of the piping, and gaskets to be used; and pressure rating of the flanges must be taken into consideration while designing. Flanges and gaskets in the pipelines should be selected as per operating conditions of pressure, temperature, and properties (corrosive, toxic, inflammable) of the fluid being handled. Special care is to be taken when dangerous (corrosive, inflammable, toxic) fluids are to be handled and appropriate safety factor shall be applied. Proper design of the piping systems will enable accurate cost estimates of piping system.

Following factors should be considered in developing design of piping system.

- Selecting the correct material of construction suitable for the properties of the fluid being handled and the worst operating conditions which may occur. Special attention should be given when dangerous fluids (toxic, inflammable, corrosive) are to be handled at high pressures and temperatures.
- Selecting the most economical pipe size to handle a given flow.
- Layout of pipeline for ease of accessibility and minimum stresses in the system.
- Provision for proper drainage of liquid/proper venting of gases before taking up maintenance work.

- Selection of the correct valves for the required operating condition and function.
- Selection of proper pipe fittings, anchors, hangers, and other supports.
- Specifying economic insulation where required as per service requirements.
- Determining proper outside colour code for identification.

Note Cost of piping increases with increasing diameter; but the cost of power required for pumping decreases as pipes of higher diameter are used. Hence it is advisable to calculate the period in which the increased cost of piping can be recovered due to reduction in cost of power required for different diameters of pipes. An efficiency of about 70 % can be assumed for the pumps; and about 85 % for the drive motor for the preliminary calculations. Cost of pipe per meter for different diameters and cost of power per KWH shall be considered as per local conditions. A payback period of about 2–2.5 years can be considered acceptable. The corresponding diameter of pipe can be selected.

5.3 Codes for Piping

Compliance to code is generally mandated by regulations imposed by regulatory and enforcement agencies. Compliance to standards is normally required by the rules of the applicable code (or the purchaser's specifications) in the country where the chemical plant is to be installed (Table 5.1).

Refer Appendix C (Tables C.1 and C.2) for ASME/ANSI and API codes for piping.

5.4 Pipe Size

ASME B36.10 and ASME B36.19 provide information for all Steel pipes and Stainless Steel pipes respectively. Pipe size is specified by:

- Nominal pipe size (NPS)
- Diameter nominal (DN)
- Schedule number (SCH)

Table 5.1 Different codes for piping system

Code	Description
ASA: American Standard Association	For pipes, flanges, and fittings
ASTM: American Society for Testing Material	Standards related to materials
AISI: American Iron and Steel Institute	Standard Steel products manual
API: American Petroleum Institute	Oil industry piping design standards
MSS: Manufacturing Standardization Society	For valves and fittings
ISO: International Organization for Standardization	Pipe, tube, and fittings standards and specifications

Nominal Pipe Size (NPS) Nominal pipe size gives diameter in inches which is related to outside diameter independent of schedule number or wall thickness. This is required to ensure interchangeability of fittings. For NPS 12 and smaller, outside pipe diameter is greater than the NPS size. For NPS 14 and larger, NPS is equal to the outside diameter.

Diameter Nominal (DN) This is close to inner pipe diameter and is the European equivalent of NPS. DN gives approximate value of pipe diameter in mm obtained by multiplying NPS by 25.

Schedule Number Schedule number gives approximate value for pipe wall thickness based on following formula.

$$\text{Schedule number} = 1000 \times P / S$$

where, P = Internal working pressure; S = Allowable stress.

For a given NPS, the outside diameter stays constant and the wall thickness increases with larger schedule number. The inside diameter will depend upon the pipe wall thickness specified by the schedule number. Larger the internal service pressure, larger the Schedule Number and vice versa. Example: The pipe of NPS one inch, has an outside diameter of 1.315 inches but the wall thickness varies with schedule number and is 0.065 inch, 0.109 inch and 0.133 inch for Schedule number 5, 10, and 40 respectively (Table 5.2).

(Reference: Perry's Chemical Engineers Handbook)

5.5 Pipe Fittings

Pipe fittings are important parts of a piping system for:

- Changing the direction of the flow.

- Elbows: For changing the direction of flow by 90° or 45°.

Table 5.2 Examples of NPS, DN and outside diameter

NPS (INCH)	DN (MM)	Outside Diameter, INCH (MM)	Schedule number (40)		SCH number (80)	
			Wall thickness (MM)	Weight of plain-end PIPE (KG/M)	Wall thickness (MM)	Weight of plain-end pipe (KG/M)
1	25	1.315 (33.40)	3.378	2.5	4.5466	3.229
2	50	2.375 (60.33)	3.912	5.4318	5.5372	7.471
4	100	4.50 (114.30)	6.02	16.058	8.5598	22.293
6	150	6.625 (168.28)	7.112	28.23	10.9728	42.517
10	250	10.75 (273.05)	9.271	60.241	12.7	81.4621
14	350	14.00 (355.60)	11.125	94.409	19.05	157.939
16	400	16.00 (406.40)	12.7	123.17	21.4376	203.298

- Tee: For making a bypass through the straight pipeline/for a branch connection.
- Reducers for connecting two pipes of different diameters.
- End caps which are internally threaded or plugs which are externally threaded are used to stop the flow from the pipes.
- Strainers in sampling lines to remove solid particles from gas/liquid lines or liquid droplets from gas lines.
- Valves to control flow of fluid.
- Non-return valves to prevent reverse flow in the pipeline.

5.6   Valves

Valve is an important piping accessory. It is a mechanical device that regulates, controls, and directs the flow of a fluid by opening, closing, or partially obstructing the flow and pressure within a system or process.

5.6.1   Mode of Operation

The operation of a valve can be carried out either manually (by hand) or mechanically. Mechanical attachments (actuators) to a valve are electrically, hydraulically, or pneumatically operated.

Types of Actuators
- Pneumatic Actuator: These need clean compressed air at sufficient pressure for operation. An air filter, air compressor, receiver for compressed air, and pressure regulator are important components of the system.

Example: Actuator for the drain valve at bottom of the process vessel is operated by compressed air made available from the air receiver. The compressor motor gets switched off at preset pressure in the receiver.

Quality of compressed air is maintained by filtering incoming air and drying the compressed air as shown in Fig. 5.1.

- Motorised valves: These have a drive motor to operate the valve spindle through a gear box. Indicators (for showing position of rotation of the internal plug/flap) and limit switches are generally provided to assist in flow control. These valves are suitable for remote operation, for large size, and when located in inaccessible/dangerous places in the plant.

- *Example: Operation of butterfly valves of large diameter gas lines at a height in a sulphuric acid plant.*
- Electro-mechanical actuators: These get electrical signal from the sensors (which are in contact with process fluid) depending on the deviation from the set point.

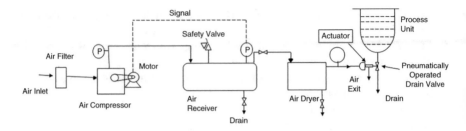

PNEUMATICALLY OPERATED DRAIN VALVE

Fig. 5.1 Schematic of pneumatically actuated valve

The electrical signal is converted to mechanical force to operate a link for taking corrective action for the flow/pressure of the process fluid. Proper calibration of the sensors and quick response of the actuating link are essential for satisfactory operations.

- *Example: Operation of valve for addition of dilution water to sulphuric acid tank as per signal received from concentration analyser.*
- Hydraulic actuators: These can be used when higher torque is required to operate the valve (example: when handling viscous fluids.) It needs high pressure oil for operation; and a return line for the oil up to the oil tank. The circulation pump installed on the oil tank supplies high pressure oil to operate the actuator for the outlet valve of the process tank. This system is costly and hence is provided for operating more than one hydraulic actuator instead of providing for a single unit.

- An oil cooler is provided to control the temperature of the oil in circulation (see Fig. 5.2).

- *Example: Operation of outlet valve for a thick viscous product from a reactor.*

Note In case of high pressure (when compressed air or oil under pressure is not being used) the compressor/oil pump motor gets a signal to trip them.

The motor gets restarted automatically when the pressure reduces.

Manual Actuators These actuators are used to operate the valves manually (example: through a chain pulley attached to a valve in an overhead pipeline).

5.6.2 Functions of Valve

- It serves to regulate the flow of liquid or gas to maintain the process parameter such as flow, pressure and level as well as it isolates the piping on equipment for maintenance without disturbing the other units.
- It helps to change the direction of flow and relieve pipe system of certain pressure.

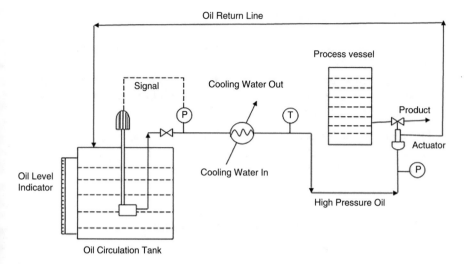

HYDRAULICALLY ACTUATED PRODUCT OUTLET VALVE

Fig. 5.2 Schematic of Hydraulically Actuated Valve

• It acts as a final control element in a control loop. Most of the valves are automatic in the modern plant. By using these valves, the variables can be manipulated to achieve the control objective.

5.6.3   Classification of Valves

• On the basis of design –This classification is based on the motion of stem.

• The available designs are

  – Linear: e.g. gate, globe, diaphragm
  – Rotary: e.g. butterfly, plug valve

• On the basis of function

  – ON/OFF or isolation valve
  – Control or moderating valve

• The third category is of non-return valves where "Swing check" and "Lift check" valves are available.

• This is used to prevent reverse flow of the fluid.

  Examples:

1. This valve does not allow back flow of reaction mixture from the reactor into its feeding line.
2. These are also used in discharge line of pumps (when more than one pump is to be run in parallel).

   Symbol of an arrow is embossed on these valves to indicate the direction of flow. Care should be taken while fitting these valves to a pipeline *as per normal direction of flow* of fluid.

Safety or Pressure Release Valve These are used in process vessels which are likely to get pressurised (due to failure of cooling water supply, due to excessive heating during reactions, etc.).

5.6.4 Types of Valves and Their Applications

Gate Valve Gate valves work by inserting a gate or a plate/wedge into the path of a flowing fluid. If the valve has a rising stem its position can be seen just by looking at the position of the stem. An indicator is also provided on the stem (of some designs) of valves to show its position.

- Gate valve is a common type of a valve where the diameter of opening through which the fluid passes is nearly same as that of the pipe. Here direction of flow does not change. The valve consists of a sliding plate or gate set in grooves. When the valve is opened the plate rises into the bonnet, completely out of the path of fluid and ceases the flow when closed by rotating the hand wheel.
- These are not recommended for controlling flow. They are used for starting/stopping flow of fluid (on-off service) or isolation of the process vessel.
- Advantages are straight through flow, low pressure drop, and can be used for fluids containing suspended solid matter.
- Disadvantages are these are slow acting, bulky, and cannot be used for throttling services.

Globe Valve The term globe refers to shape of the style of the valve body. In this the orifice is arranged perpendicular or angular to the axis of flow.

- The closing or opening is accomplished by a disc or plug, attached to the stem which may be rotated to open or close the circular orifice. Fluid flow through valve must make approach to the 270° turn to reach the orifice. After passing through the orifice, it turns another 270° to the original flow.
- Globe valves are used primarily in situations where throttling of the liquid is required. By simply rotating the hand wheel, the rate at which the liquid flows through the valve can be adjusted to any desired level. Having the valve seat parallel to the line of flow is an important feature of the globe valve. This feature makes the globe valve efficient when throttling liquids while minimising disc and seat erosion.

- Advantages are that they can be primarily used for throttling the flow of liquid, and used for controlling process fluid under variety of operating conditions such as high temperature, high pressure, viscosity, and low pressure.
- Disadvantage is high head loss.

Diaphragm Valve These valves are employed where the valve mechanism must be isolated from the fluid or where closure is required when the fluid contains suspended solids. In this type of valve, the diaphragm acts as a closing medium as well as sealing gasket. By moving diaphragm close to or away from weir, the flow is controlled.

- Diaphragm valves (or membrane valves) consists of a valve body with two or more ports, a diaphragm, and a "weir or saddle" or seat upon which the diaphragm closes the valve. The valve body may be constructed from plastic, metal, or other materials depending on the intended use.
- There are two main categories of diaphragm valves: one type seals over a "weir" (saddle) and the other (sometimes called a "full bore or straight-way" valve) seals over a seat.
- An air compressor is required for the valve actuation. When air pressure is applied on diaphragm; the diaphragm moves towards the weir and the flow decreases. When the pressure is withdrawn, the diaphragm moves away from the weir.
- These have a flexible rubber/teflon/teflon-coated diaphragms which can be pressed on the seat to control the fluid flow. The operating stem is above the diaphragm.
- These are used in pulp and paper industries; pharmaceutical plants and waste water treatment plants.
- The diaphragm action can be actuated manually or with a pneumatic actuator (the valve body base remains the same).

Butterfly Valve The elements of this valve are valve body, disc, and shaft. The discs are generally welded or bolted to the shaft.

- The disc can be rotated around shaft by rotating the shaft either manually or automatically by attaching electrical, pneumatic, or hydraulic motor drive to the shaft. The 90° rotation of shaft will convert the valve from fully open position to fully close position.
- When in the fully open position, the disc is parallel to the flow and in fully closed position the disc is perpendicular to the flow.
- Butterfly valves are generally used in furnaces, dampers, gas lines; water lines, etc.

Ball Valve Ball valve consists of spherical plug and a valve body. The rotation of the spherical plug in the valve body controls the flow. Major components of the ball valve are the body, spherical plug, and seats. Ball valves are made in three general patterns: venturi port, full port, and reduced port. The full-port valve has an inside diameter equal to the inside diameter of the pipe. In the venturi and reduced-port styles, the port is generally one pipe size smaller than the line size. Stem sealing is accomplished by bolted packing glands and O-ring seals.

- Their advantages are that they are quick acting, have easy operation and have straight through flow.
- Ball valves can be one-piece, two-piece or three-piece valves. In one-piece ball valve there is one solid cast body. Two-piece ball valves have two pieces - one piece has one end connection and the body and the second piece fits into the first. Three-piece ball valves have one main body and two pipe connectors which are threaded or welded to pipe. The main body may be easily removed for cleaning or repair without removing the pipe connectors.

Needle Valve These valves have a handle connected to a plunger or stem. The stem moves up or down when handle is turned. As the stem moves down, the tapered pointed tip comes into contact with valve seat to seal the orifice.

- These are used for low flow rate such as fine adjustment of flow of fluids (e.g. fuel oil flow to burner).

Pinch Valve These valves have a flexible tube in the body. The flow through the tube can be pinched and closed by the pressing a plug on it.

- This valve is used for control of flow at lower pressure. Not suitable for abrasive slurry or liquid containing hard suspended particles.

5.6.5 Material of Construction of Valves

Commonly used material of construction for valves are cast iron, brass, stainless steel, polypropylene, etc. selected on the basis of properties of fluids handled; operating conditions of pressure, temperature, pH, and presence of suspended abrasive particles; and cost of valve.

Lined valves are used for corrosive liquids. Perfluoroalkoxy (PFA) and polytetrafluoroethylene (PTFE) (teflon) lining/neoprene rubber lining are used for highly corrosive liquids due to their excellent mechanical strength and stress-cracking resistance properties.

Polypropylene (PP) valves find application for regulating the flow of viscous and non-viscous fluids and are used in paper, sugar, fertilizer, chemical, nuclear plants, refineries, and petroleum industries.

5.6.6 End Connections for Valves

- For medium and larger diameter pipes, flanged connections are used. The flanges and gaskets are selected based on the properties of fluids and operating conditions.

- For small diameter of piping, threaded connections are used for handling liquids at lower pressure. These are prone to leakages from threaded connection with piping.

5.6.7   Basis for Selection of Valves

- Purpose to use the valve i.e. for control, isolation, or shut-off.
- Maximum fluid flow rate.
- Characteristics of valve.
- Nature of fluid flowing through valve e.g. abrasive, corrosive, and viscous.
- Pressure drop in valve.
- Fluid temperature through valve whether cryogenic (Freezing mixture).
- Presence of suspended solids.
- Cost

5.6.8   Valve Characteristics

The relation between the flow through valve and the valve stem position is called the valve characteristics. All control valves have an inherent flow characteristic that defines the relationship between "valve opening" and flow rate under constant pressure conditions. Valve opening refers to the relative position of the valve plug to its closed position against the valve seat.

The types of valve characteristic can be defined in terms of "sensitivity" of valve, which is the fractional change in flow to fractional change in lift position for fixed upstream and downstream pressure. In terms of valve characteristics, valves can be divided into three types (Fig. 5.3):

- **Linear**: Sensitivity is constant and there is linear relation between flow and valve opening at constant pressure drop i.e. flow is directly proportional to valve opening. For linear flow characteristic valve, for equal increment in valve travel, there is equal increment of flow.

- These valves are often used for liquid level or flow control and for processes where pressure drop across valve is expected to remain constant.

$$\text{i.e. } \frac{Q}{Q_{max}} = \frac{S}{S_{max}}$$

- where, Q = Flow at constant pressure drop; S = Valve opening.
- **Equal percentage**: Sensitivity (or slope) increases with flow i.e. flow increases exponentially with valve travel. For equal percentage valve, equal increment in

valve travel produces equal percentage changes in flow at a constant pressure drop based on the flow just before the stem position is changed. Hence, most important property of such valves is a constant percentage rate of flow change at a constant pressure drop per unit change in the valve opening through the major part of the stroke. These types of valves are commonly used for pressure and temperature control applications and are most suitable for processes where a high variation in pressure drop is expected. The general equation is:

$$Q = Q_{min}.R^{\frac{S}{S_{max}}}$$

- where, Q = Flow at constant pressure drop; S = Valve opening
- Rangeability R is defined as the ratio of maximum flow to minimum controllable flow (Q_{max}/Q_{min}).

$$R = \text{Rangeability} = \frac{Q_{max}}{Q_{min}}$$

- **Quick opening**: For this type, sensitivity (or slope) decreases with flow. For these valves the flow v/s percentage valve opening characteristic relationship is approximately linear initially with very steep slope at a low percentage of stem

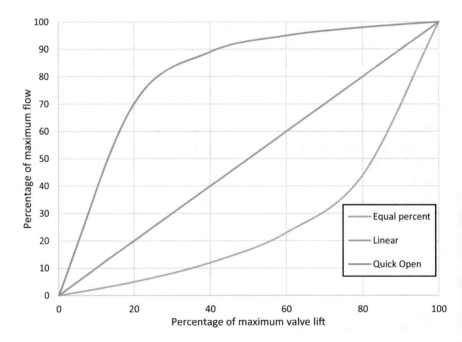

Fig. 5.3 Valve characteristics

travel i.e. a small initial change in stem travel produces a large increment in flow (almost maximum flow is attained for low percentage of maximum stem lift).

- These are normally used for "on-off" applications and processes where instant large flow is required particularly in cooling water and or safety systems.

5.7   Pipe Joints

Pipe joints are essential to join the pipe to other pipes or other components. Ideal pipe joints do not change any dimension of the flow passage or direction of flow. Also, it should require minimum labor to disassemble. The method used to join pieces of pipe depends mainly on material properties apart from wall thickness and operating conditions of pressure and temperature.

Commonly used pipe joints are as follows:

Welded joints: Welded joints must be ground smooth internally if the diameter is large to reduce friction loss, vortex formation, etc. Welded joints are of following types:

- Butt welded: The most widely used welded joint in piping system is the butt weld joints. In all ductile pipe metals which can be welded, pipes, elbow, tee, and valves are available with ends prepared for butt welding and in all sizes.
- Socket joint: Plain end pipe used for socket weld joints is available in all sizes. The wall thickness of the socket must be equal to or greater than 1.25 times the minimum wall thickness of pipe. This joint is less resistant to bending stress than the butt welded.
- Branch welded: Branch weld eliminate the purchase of tees.

Flanged joints: It is one the most widely used pipe joint for size larger than (50 mm) where separation of the connected pipes may be required (say, for cleaning inside). The flanges are connected by means of bolts. Flanges have been standardised for pressure up to 2 N/mm². The flanged faces are machined to ensure correct alignment of the pipes. The joints are to be made leak proof by providing gaskets of appropriate thickness and material. Flange thickness, bolt sizes, bolt circle diameter shall be selected as per service conditions, for the fluid properties, operating pressure and temperature, etc.

Socket and spigot joint: These types of joints are mainly used where the pipes are to be used underground. The joint being flexible adapts itself to small changes in level due to the settlement of earth. The spigot end of one pipe enters the socket end (which has a greater diameter) of the other. The assembly is then sealed from outside by suitable cement.

Expansion joints: Piping carrying steam at high pressure is usually made to permit longitudinal expansion or contraction of metal due to variation in temperature. Examples of expansion joints are corrugated joints, loop joints, gland, and stuffing box expansion joint.

Important Precaution Full covers are to be provided on all flanged joints of pipelines carrying corrosive or dangerous fluids so that any leaking joint will not injure persons below.

5.8 Colour Coding for Piping

Various fluids (gases and liquids) transported by pipes at times can be toxic. In order to identify the material being transported, it is necessary to have proper standard colour coding for pipes. Different colours are used to indicate the type of fluid being carried. This would reduce possibility of mistakes in identification and accidents due to wrong identification of pipelines during emergency situation.

There are national and international standards that provide the guidelines for uniform colour coding for piping systems in industries.

Codes used for colour coding

- ASME A13.1: Scheme for the Identification of Piping Systems

- This American National standard helps in identification of hazardous materials conveyed in piping systems and their hazards when released in the environment.
- IS 2379: Colour code for identification of pipelines.
- As per this Indian Standard (IS 2379), ground colour indicates the basic nature of fluid carried and shall be applied throughout the length and colour bands are superimposed on ground colour at different locations to distinguish one fluid from another but belonging to the same group.
- *(Refer Appendix C, Table C.3: Piping colour code as per IS 2379)*
- BS 1710: Specification for Identification of Pipelines and Services.

- Colour coding as per BS 1710 uses two types colour coding to identify the content of pipe and hazard.

- Base colour—Base colours are used to indicate the content inside the pipe.

- Safety colours—These colours are used as band colours that are applied in conjunction

5.9 Important Terminology to Be Provided with Piping and Instrumentation Diagram

- Anchor: Point where piping is fixed is called an anchor.
- Guide: Device controlling the direction of piping movement. Unlike the pipe anchor, which is welded to the pipe and steel structure, guide allows pipe to slide lengthwise between two angle shapes.
- Bleed/Vent: A small valve provided to bleed off liquid.

- Blind: A plate sized to be inserted in a flanged joint to isolate a portion of the system.
- Block valve: Any type of valve that is used to close the flow, such as flow through a pipeline or out of a tank, such as gate valve, ball valve, or plug valves.
- Bottom of pipe (B.O.P): This is used for pipe support location and to know the elevation at which the pipe wall ends/starts.
- Directions: North direction of the plant as on plot plan and other drawings will serve for orientation of equipment.
- Drip leg: A vertical section of pipe located in horizontal piping to deflect and catch condensate.
- Gradient: The successive drop in elevation of piping to ensure gravity flow and drainage.
- The designation line number: It is a number and a symbol appearing on piping drawing which indicate the pipe according to size, process liquid, and specification.
- Pipe support (P.S.) is used to indicate the type of pipe support.

5.10 Piping Stresses

Types of stresses induced in piping are:

- Stress caused by internal and external pressure.
- Stresses remaining in pipe wall after fabrication or erection.
- Stresses caused by temperature changes in process fluid.
- Stresses due to weight of pipe insulation and pipe fittings.
- Stresses due to vibration.

Methods for Relieving Expansion Stresses
In most piping systems stresses are caused by temperature changes in process fluid in the pipe. Some means must be provided for reducing these stresses. Provision to be made in piping carrying steam at high pressure is to permit longitudinal expansion or contraction due to variation in temperature.

The piping must be made flexible to avoid these stresses. This can be accomplished by the use of expansion joints, providing flexibility in the piping layout or by cold springing.

- Expansion joints: These can be metallic or made of natural or synthetic elastomers, rubber, etc. They help to safely absorb dimensional changes of pipes due to temperature changes, vibrations caused by rotating machinery, misalignment during installation, shock effects due to water hammer, or pump cavitation. These shall be selected as per operating conditions and properties of fluids.

There are different types of expansion joints. Examples: corrugated joints, loop joints, gland, and stuffing box expansion joint.

- Flexibility through layout: Piping is rarely used in a straight line from one fixed connection point to another. Each time piping changes direction and is free to move at the point of change, it makes the system more flexible. Provide S- shaped bends/pieces wherever possible.
- Cold springing: Forces generated by expansion are caused by movement of piping as its dimensions change from atmospheric to operating condition. One method for decreasing the final intensity of the stress in the hot condition is to cut the pipe short and spring (joint) it into place while cold. This springing produces tensile stress on the system which changes to a compression stress in the expanded hot condition. This technique of installation is called cold springing. *This cannot be used for cast iron/glass lined piping or those carrying dangerous fluids.*

5.11 Underground Piping

Advantages:

- Laying a pipeline in a tunnel reduce the heat losses and insulation cost.
- It does not occupy overhead space.
- Protects the piping from mechanical damage.

Disadvantages:

- Leakages would be difficult to locate underground.
- This would require frequent inspection.
- Would make repairs difficult when piping must pass under certain obstacle. It should be installed in such a manner so as to permit it to be withdrawn readily for replacement.
- Underground piping is also subjected to the usual underground corrosion.

5.12 Piping Layout

Plant engineers and project managers must check the P & ID by carrying out HAZID and HAZOP studies for piping layout. The most economical layout usually results in a lane or shadow design in which all north-south piping is located at one elevation and all east-west piping at another. This allows for future expansions or modification without interference. The key lines and large expensive piping are positioned first for shortest and most direct route.

After all piping and auxiliaries are installed on model, different suggestions from representatives of design groups are taken for any corrections.

Layout design must meet following requirements:

- Easy and economical installation (optimisation of initial cost balanced against cost of energy for transport of fluids).
- Shall not obstruct movement of personnel and materials.
- Operational ease and accessibility.
- Ease of draining/venting inside fluids.
- Minimum possibility of formation of idle pockets.
- Ease of inspection, maintenance, and replacement.
- Protection of piping system from physical or thermal shocks.
- Adequate independent supports without hampering accessibility to equipment.
- Compliance to safety codes followed for specific piping.

Piping Layout Rules   Keep piping above ground, if possible, for the ease of installation and repairing. Avoid overhead layout for structural reasons.

- Provide fire stops and drains for underground piping.
- Allow space for extra piping.
- Allow space for other utilities.
- Considerations for positioning of valves in the layout are:

  All valves should be readily accessible from floor level.
  Valves on lines discharging in open tanks should be located so that the operator is not exposed to splashing or fumes.
  Locate emerging dump valves and discharge line at safe distance from the operating area (outside of the process building if possible).

Series and Branched pipes   Total flow through pipes in series is same and head loss is the summation of head losses in each pipe. For parallel pipes, the total flow is summation of flow in each pipe and head loss across each pipe is same.

5.13   Pressure Drop in Pipelines

Pump power and head can be determined by finding hydrostatic pressure difference and pressure drop in pipeline. Pressure drop occurs due to skin friction in pipe called major losses and due to sudden expansion, contraction, and pipe fittings such as valves, bends, tees, and elbows called minor losses. The total pressure loss in piping systems is the sum of all.

Table 5.3. gives relations to find major and minor losses.

Sample Problem   Develop a family of curves of pressure drop along standard steel pipe of length 300 m for pipe diameters of 400 mm, 450 mm, and 500 mm for different flowrates of 0.4 m^3/s and 0.5 m^3/s and 0.6 m^3/s , considering fluid flowing as water.

(Refer Appendix "Sample Calculation 1" for detailed calculations and Python code)

We know that pressure drop due to skin friction in pipes is given by

$$\Delta P = f \frac{L \rho \overline{V}^2}{2D}$$

where, ΔP = Pressure drop; L = Length of pipe; D = Inside diameter of pipe; V = Velocity of flow; ρ = Density of fluid; f = Friction factor.

Calculated values of pressure drop for different flow rates and pipe diameters have been plotted below (Fig. 5.4).

Non-circular Conduits For non-circular pipes, equivalent diameter (D_e) is found, which is four times hydraulic radius.

$$D_e = 4R_h$$

$$R_h = \frac{\text{Crosssectionalareaofflow}}{\text{WettedPerimeter}}$$

R_h = hydraulic radius

Wetted perimeter is the perimeter of the cross section that is in contact with the flowing fluid.

Practical Way of Finding Pressure Drop Across Pipe Fittings Pressure drop in pipes with fittings can be calculated by following methods:

Equivalent Length Equivalent length of a valve or fitting is the length of straight pipe of uniform diameter that for same flow rate would have same frictional resistance and produce same head loss as the frictional loss due to valve or fitting. For pipe fittings this term is used for easy calculation of pressure drop. Pressure drop can be calculated using Darcy-Weisbach equation after finding the equivalent length corresponding to the pipe fitting.

Velocity Head Velocity head is used to represent kinetic energy of fluid per unit weight and has physical dimension of length.

Overall head loss, h_m is commonly reported in velocity heads, i.e. $\frac{V^2}{2g}$ as

$$h_m = K_L \frac{V^2}{2g}$$

h_m = Overall head loss; K_L = Overall loss coefficient.

The total loss can be summed as

$$h_m = \left(4f \frac{1}{D} + K_c + K_{ex} + K_{ff} + K_e \right) \frac{\overline{V}^2}{2g}$$

Table 5.3 Formulae for pressure drop calculations

Type of loss	Pressure drop calculation	
Hydrostatic pressure difference	P = Hydrostatic pressure difference = $\rho g Z$ Z = Vertical elevation change g = Acceleration due to gravity ρ = Density of fluid	
Friction Pressure Loss by Darcy Weisbach equation	$h_{fs} = f\dfrac{L\bar{V}^2}{2gD}$ h_{fs} = head loss in m, ft f = Darcy Weisbach friction factor L = Length of pipe $\bar{V}$ = Average flow velocity of fluid D = Inner pipe diameter, $\Delta P = f\dfrac{L\rho\bar{V}^2}{2D}$ ΔP = Pressure drop ρ = Density of fluid	For laminar flow, (Re < 2100) $f = 64/\text{Re}$ Re = Reynolds number $= \dfrac{\rho v D}{\mu}$ μ = Viscosity of fluid ρ = Density of fluid Hagen Poiseuille relation, $\Delta P = \dfrac{32\Delta L\bar{V}\mu}{D^2}$
		For turbulent flow, – For smooth pipes Blasius equation $f_D = \dfrac{0.316}{\text{Re}^{1/4}}$ for $3000 < \text{Re} < 10^5$ – For rough pipes, Colebrook equation $\dfrac{1}{\sqrt{f}} = -2\log\left(\dfrac{\epsilon/D}{3.7} + \dfrac{2.51}{\text{Re}\sqrt{f}}\right)$ ϵ/D = relative roughness which is ratio of height of roughness of the pipe to diameter of pipe
Losses in pipes due to fittings such as bends, tees, elbows, unions, valves	Losses due to pipe fittings $h_{ff} = K_f\dfrac{\bar{V}^2}{2g}$ $\bar{V}$ = Average velocity in pipe leading to fitting. K_f = Loss coefficients for standard threaded pipe fittings	

Type of loss	Pressure drop calculation
Losses in pipes due to sudden expansions, contractions	Loss due to sudden expansion $$h_{fex} = K_{ex}\frac{V_1}{2g}$$ $$K_{ex} = 1 - \frac{d^2}{D^2}^2$$ d = diameter of smaller pipe D = diameter of larger pipe V_1 = Velocity at inlet K_{ex} = Expansion loss coefficient
	Loss due to sudden contraction $$h_c = K_c\frac{\overline{V}_2^2}{2g}$$ $\overline{V}_2$ = Average velocity in smaller downstream section K_c = Contraction loss coefficient For laminar flow $K_c = 0.1$ $$K_c = 0.4\left(1 - \frac{A_2}{A_1}\right)$$ For turbulent flow where A_1 and A_2 are cross-sectional areas of upstream and downstream pipes respectively
Entrance and Exit Losses:	$$h_f = K_e\frac{V^2}{2g}$$ K_e = Entrance or exit loss coefficient Depending on the pipe geometry at exit and entry value of loss coefficient is taken as For sharp-edged inlet, $K_{e} = 0.5$ For rounded inlet, $K_e = 0.04$

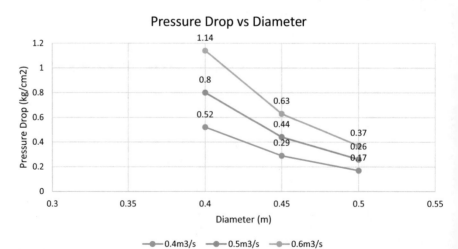

Fig. 5.4 Plot of pressure drop vs. pipe diameters at different flow rates

Table 5.4 Pressure loss in pipe fittings and valves

Pipe fitting	K, number of velocity heads	Number of equivalent pipe diameters
45° Standard elbow	0.35	15
90° Standard radius elbow	0.6-0.8	30-40
90° Square elbow	1.5	75
Tee entry from leg	1.2	60
Tee entry into leg	1.8	90
Gate valve fully open	0.15	7.5
Gate valve half open	4	200
Globe valve fully open	6	300

Velocity head and equivalent length for various pipe fittings are available in liteature (Table 5.4).

These are some typical values taken from Coulson and Richardson's Chemical Engineering Design, R.K.Sinnott, Fourth Edition Vol.6 pg. 204. The reader may please refer to this book for more information

5.14   Material of Construction for Pipes

- Mild steel: These are used for pipes for water, steam and condensate; concentrated sulphuric acid at ambient temperature and low velocities, oleums of high strength, organic chemicals, petroleum products, alkali, and various gases. Piping systems consisting of these pipes are easy to fabricate. These pipes can handle fluids under pressure.
- Cast Iron: These pipes have good corrosion resistance and can handle concentrated sulphuric acid at 70–75 °C. However, cast iron pipes cannot be used for handling oleums.
- CI pipes should be properly supported as they cannot take tensile loads. Never over tighten the flanges in CI pipelines (specially to bring together adjacent pipes if there is a slight gap. Insert a CI/stainless steel spacer in the gap and then tighten the joint normally.
- Stainless steel: They have very good resistance to many chemicals except HCl and chlorides. These pipes can handle fluids under pressure. These pipes are costly compared to MS pipes.
- Glass: These pipes are used for handling pure chemicals as they do not cause contamination. They are brittle and costly. They can get damaged due to tensile loads or vibration; hence glass pipelines need very good supports and should be firmly clamped (after providing rubber pads). Teflon gaskets are to be used.
- Teflon lined stainless steel external braided hose pipes: They are used for handling very corrosive chemicals like liquid SO_3.
- The internal lining of teflon imparts corrosion resistance while the outside stainless steel gives mechanical strength.

- Polypropylene piping: They are used for corrosive chemicals at ambient temperature/below 70 °C. However, they are not advisable for handling concentrated sulphuric acid or oleums.
- Rubber hose pipes are used for water, compressed air, dilute alkali, and oil. These shall be tested at 1.5 times the operating pressure before taking in to use.
- PVC pipes are used for handling waste waters and drainage services. But they can get affected by long exposures to strong sunlight or high ambient temperatures.
- **Important: Please select the material for pipes to suit the operating conditions and properties of fluids**.

5.15 Pumping System

Pumps convert mechanical energy supplied to it from external source into pressure energy which is used to move liquid from one point to another or lift it from a lower level to higher level. Pumps are used for handling liquids, solutions, or slurries.

5.15.1 Classification of Pumps

Pumps are classified based on their functions and also the working principle by which energy is added to the liquid.

Positive displacement pump: In a positive displacement pump, energy is periodically added to the liquid by direct application of a force to one or more movable volumes of liquid. This causes an increase in pressure up to the value required to move the liquid through the discharge line. The energy addition is periodic (i.e. not continuous) and there is a direct application of force to the liquid. Reciprocating piston or plunger pumps are examples of positive displacement pumps. Definite volume of liquid enters through inlet and is trapped in a chamber followed by exit through discharge at higher pressure. As the piston or plunger moves back and forth in the cylinder, it exerts a force directly on the liquid, which causes an increase in the liquid pressure. Rotary pumps are also positive displacement pumps and in this as the gears rotate, fluid flow occurs through casing and discharge created.

Rotodynamic/Pressure pump: These pumps have a rotating element called impeller through which as the liquid passes, its angular momentum changes due to which pressure energy of the liquid is increased. Unlike positive displacement pump they do not push the liquid. In this velocity of liquid is increased by continuously adding energy to it (Table 5.5).

5.15.2 Pump Selection

Type of pump is selected based on the flow rate, pressure/discharge head of liquid to be pumped, and liquid properties such as viscosity, specific gravity, temperature, corrosive nature, vapor pressure, solids present, and position of installation. Material of construction of pump should be compatible to handle the liquid and the environment in which it is to be installed. Pump power can be calculated by knowing the inlet and outlet pipe size, fittings and pressure losses, flow rate, and total discharge head. Net positive suction head available should be calculated by knowing pump inlet pressure and liquid vapor pressure.

Typical Considerations for Pump Selection
The following shall be considered while selecting a pump for the plant.

- Capacity (flow rate) required (m^3/h).
- Normal operating discharge required. (kg/cm^2).
- Maximum pressure the pump can develop. This should not damage process piping/units, etc.
- Operating temperature (normal and maximum).
- Properties of the fluid such as density and viscosity as operating temperature.
- Presence of suspended solids in the fluid. Selection of impeller depends on this.
- Corrosive, inflammable, or toxic properties of the fluid. The pump should be provided with mechanical shaft seal in such cases. Conventional gland packing or water cooled gland packing can be used for harmless liquids.
- Wetted parts of the pump like impeller, shaft, and wearing ring should be of corrosion-resistant materials such as stainless steel, Alloy—20, Hastelloy, etc.
- Whether a net positive suction head is available at the given location. Submerged pump may be selected if liquid is to be sucked from very deep underground tanks.
- Location of installation i.e. whether indoor or outdoor.
- Position of installation, whether outside a process tank or inside the tank.
- Operating speed of pump and drive motor HP required (to be informed by the manufacturer).
- Accessories like base plate, coupling halves, guard, and foundation bolts shall be provided by the manufacturer.

  Commonly used pumps have been discussed here.

Table 5.5 Classification of pumps

Type of pump	Some typical examples
Positive displacement pumps	Reciprocating Pumps: e.g. plunger pump, piston pump, diaphragm pump
	Rotary pumps: e.g. gear pump, lobe pump, screw pump
Rotodynamic/Pressure pump	Centrifugal pump
	Axial flow pump
	Radial pump

5.15.2.1 Positive Displacement Pumps

Reciprocating Pumps
In this type of pump liquid is sucked and pushed or displaced due to thrust exerted on it by a moving member i.e. piston/plunger resulting in displacing/lifting of liquid to desired height. It consists of a piston which moves to and fro in close fitting cylinder as the crank rotates. The cylinder is connected to suction and delivery pipes with suitable valves. A definite volume of liquid is trapped in the chamber from suction pipe and the movement of piston discharges liquid at high pressure.

Classification of Reciprocating Pumps
Based on liquid contact with piston sides:
Single Acting:

- Liquid in contact with one side of piston.
- Has one suction and one delivery pipe.
- One complete revolution of crank produces two strokes - suction stroke and delivery stroke.

Double Acting

- Liquid in contact with both sides of piston.
- Has two suction and two delivery pipes.
- During each stroke when suction take place on one side of piston, the other side delivers the liquid.
- One complete revolution of the crank produces two suction strokes and two delivery strokes.

Based on number of cylinders provided:

- Single cylinder pump: This maybe either single acting or double acting.
- Double cylinder pump: This has two single acting cylinders each equipped with one suction and one delivery pipe with appropriate valves and separate piston for each of the cylinders.
- Triple cylinder pump: This has three single acting cylinders each equipped with one suction and one delivery pipe with appropriate valves and separate piston.
- Duplex double acting pump: This is formed by combining either two double acting single cylinder pumps or double acting double cylinder pumps.

Characteristics of Single Acting and Double Acting Reciprocating Pumps
Plot of pump discharge vs. crank angle is the characteristic for reciprocating pump. The volume of liquid drawn by piston into cylinder varies sinusoidally with angle.

In case of single acting pump as the piston moves from left to the right (i.e., as the crank angle varies from 0° to 180°), there is only suction and no discharge, then as crank angle varies from 180° to 360°, there is discharge of the liquid only. Thus, during first half revolution of crank there is suction stroke and during the next half revolution, there is delivery stroke. Thus, discharge through the pump is alternate pulsating type. Due to this type of discharge, the liquid flow will be non-uniform in the entire piping system adjacent to the pump and is subjected to fluctuating pressures. It may cause jerks in the piping. This is the limitation of the single acting pump.

In case of double acting pump, during each stroke, when suction takes place on one side of the piston, the other side delivers the liquid and thus in one complete revolution (0–360°) of crank there are two suction strokes and two delivery strokes. Efficient method to reduce the pulsating discharge (flow) by the single acting and double acting reciprocating pump is to use multiplex pump.

Discharge (Q) v/s crank angle (θ) for single acting, double acting and ideal flow have been shown in Figs. 5.5 , 5.6 and 5.7 respectively.

Problems with Reciprocating Pumps
Pump does not deliver or delivers too less
   Typical reasons

- Strainer in suction line is clogged.
- Suction line is clogged/has too small diameter.
- Air ingress through leaking suction line.
- Liquid vapor lock in suction line due to high temperature.
- Clogged discharge line.
- Discharge valve remaining closed due to any reason.
- Malfunctioning of valves in suction and/or discharge side.
- Internal pressure relief valve stuck in partially or fully open position (if it is set to open at a lower pressure than discharge pressure). It is also possible that the spring for the valve is damaged.

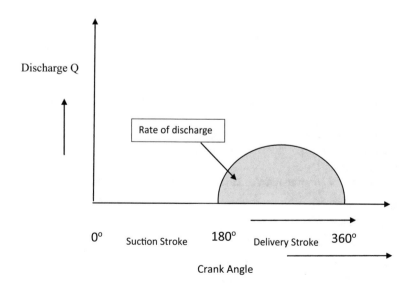

Fig. 5.5 Discharge of single acting pump

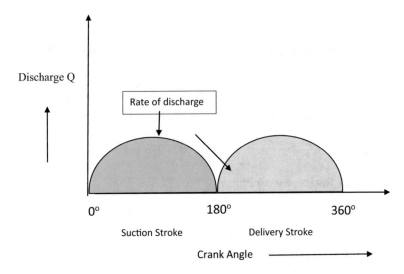

Fig. 5.6 Discharge of double acting pump

Fig. 5.7 Ideal pump
discharge

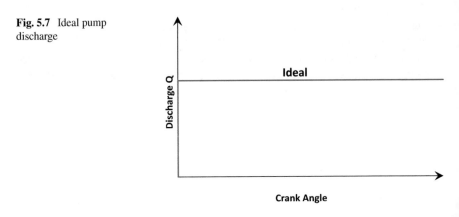

- Incorrect setting of stroke length of reciprocating pumps can result in low rate of discharge.
- Internal safety relief valve partially open.
- Mechanical damage of crank connection.
- Leaking piston ring.
- Drive motor of low capacity/incorrect lower speed/loose drive belts.

Too many vibrations
Typical reasons

- Piping inadequately supported.
- Loose foundation bolts or base plate.
- Improper alignment with drive motor.
- Loose coupling/damaged coupling with motor.

- Loose/damaged connection between connecting rod and plunger/piston.
- Operating above maximum discharge pressure permitted by manufacturer.
- Internal (spring loaded) relief valve opening frequently.
- Air ingress in suction side.
- Dampening pot located far away from pump delivery nozzle.

Drive motor is overloaded/tripping frequently
Typical reasons

- Too viscous liquid (liquid can become very viscous in winter due to non-functioning of steam traps provided on heaters/jackets for liquid lines).
- Pump being run at higher speed than recommended by manufacturer.
- Excessive discharge line pressure.
- Clogged discharge line.
- Closed/throttled valve in discharge line.
- Over-tightened gland packings.
- Trip relay set at lower current value.
- Drive motor of lower HP provided.

Too much fluctuations in discharge flow
Typical reasons

- Frequent opening of pressure safety valve PSV installed in the recirculation line. This can recycle the liquid back to suction and cause reduction in forward flow, fluctuations in pressure, and vibrations. It is possible due to lower setting of the PSV or damage of the loading spring.
- Pulsation dampening pot not provided at all.
- Pulsation dampening pot is of too small capacity.
- Pulsation dampening pot is provided far away from pump outlet (delivery side).
- Rupture of the diaphragm in diaphragm type reciprocating pumps.
- Rupture of the bladder holding inert fluids/gases in the pulsation dampening pot.

Pulsation (Pressure Fluctuation)
This is caused by the mass of a liquid being accelerated and decelerated by a pumping action, or the repeated opening and closing of a valve. It can cause unsteady flow or vibrations in the piping system. Reciprocating pumps (diaphragm or piston pumps) can create pulsation due to the reciprocating strokes. This can be minimised by providing a pulsation dampener in the delivery line as close to the pump outlet. It also reduces the energy required to pump the liquid over long piping.

Procurement of pulsation dampener
The pulsation dampener shall be designed as a pressure vessel and tested accordingly before installation. The internal bladder should be able to withstand corrosive conditions when handling corrosive liquids. It shall be filled with nitrogen at a pressure appropriate for providing the dampening action in the operating pressure range.

Following information shall be available:

- Type of pump (simplex, duplex, triplex, gear).
- Properties of liquid being pumped (density, viscosity, pH).

- Flow rate of liquid.
- Maximum discharge pressure of pump.
- Operating pressure and temperature in the discharge lines.
- Delivery pipe length, diameter, and maximum pressure rating.

Surging During Pumping Operations

Pulsation during pumping is predictable as pressure changes in piping could be expected at regular intervals due to operation of pump delivery valves; known temperature of fluids; known levels in tanks; etc. Pulsation dampeners can take care of such matters.

However sudden larger pressure and volume fluctuations can occur due to tripping/restart of pumps; emptying out of suction side tank; or sudden closure of valves in piping. These can cause much larger vibrations which may not be taken care of by the pulsation dampener. Such unpredictable pressure/volume changes are known as surges. These can be minimized by provision of surge suppressors (which can have similar design) at various points throughout the fluid system. A detailed study of the piping and pumping systems will have to be carried out for these.

Piston and Plunger Pumps

These are positive displacement type (reciprocating type pumps) which have a plunger or a piston which moves in a reciprocating manner through a cylindrical chamber. This causes drawing of fluid into the pump and delivering the fluid out of the pump.

Plunger Pump Here the high-pressure seal is stationary and a smooth cylindrical plunger moves through the seal. Hence, they can be used at higher pressures in ranges of 30—1200 kg/cm^2. This type of pump is often used to transfer municipal and industrial sewage as well as for injecting treated water into oil wells. Plunger pumps are also used for injection of corrosion inhibitor at controlled rate in systems operating at high pressure.

Piston pump is a type of positive displacement pump where the high-pressure seal moves with the reciprocating piston. Piston pumps can pump viscous fluids and those having suspended solid particles. Piston pumps are used for lower discharge pressure applications as compared to plunger pumps. However, the seal wears out earlier due to the movement.

Axial piston pumps contain a number of pistons attached to a cylindrical block.

Radial piston pumps contain pistons arranged radially from the cylindrical block. A drive shaft rotates this cylindrical block which pushes the pistons delivering the fluid.

Considerations for Selection

- Flow rate required.
- Maximum discharge pressure
- Operating temperature
- Properties of the fluid (viscosity, density, pH etc.)
- Concentration and abrasive properties of suspended solids

- Volume/stroke
- Power required to operate
- Space required for installation, etc.

Drive

These pumps have a reciprocating piston/plunger which are operated by a crank mechanism. It is driven by electric motors, hydraulic, or have steam-operated drive.

Flow Rate and Power Required

$$\text{Flowrate} = (\text{area of the plunger / piston}) \times (\text{stroke length}) \times$$
$$(\text{number of strokes / min}) \times (\text{number of pistons / plungers})$$

Power required depends directly on (discharge pressure) × (flow rate)

Materials of Construction

Wetted parts of the pump (piston/plunger, valves) should not get corroded and eroded by the fluid being handled. The casing should be able to withstand the operating conditions of pressure, temperature, pH, presence of suspended solids, etc.

Stainless steel, bronze, brass, nickel alloy, etc. are some of the materials of construction.

Diaphragm Pumps

Different varieties of diaphragm pumps are where:

1. The fluid to be pumped is on one side of the sealed diaphragm. Compressed air or hydraulic fluid is used to push the diaphragm and is on the other side. Non-return valves are provided in the suction and delivery side of the pump.
2. The fluid to be pumped is on both sides of the diaphragm which is not sealed.
3. The diaphragm is operated by an actuator which in turn is moved by a crank driven by suitable mechanical means. The fluid to be pumped is on one side of the sealed diaphragm while the other side is open to air.

Working of the Pump

Fluid to be pumped enters the pump when the diaphragm moves up and reduces the pressure inside. The pressure increases when the diaphragm moves down and the fluid is pushed out from the pump.

The diaphragm works as a moving seal which does not leak and has very little friction. The pump can operate over a considerable range of pressures and temperatures when the right type of material is used for the diaphragm.

Accessories Required

Air compressor, receiver, pressure regulator, piping, etc. for operating the diaphragm. Hydraulic power systems can also be used where greater force is required for operation.

General

• Capacity of the pump depends on the effective working diameter of the diaphragm and its stroke length.

- These pumps can handle corrosive, abrasive, toxic, slurries in explosive manufacture and flammable liquids where it is not advisable to use electric motor for operation.
- They can also be used for handling acids and caustic solutions; and for wastewater.

If the delivery rate of the diaphragm is low the following are the likely reasons.

- Low air supply pressure.
- Valve in air supply line is throttled.
- Suction line or discharge line of diaphragm pump is clogged.
- Liquid is too viscous.
- High concentration of suspended solids.
- Diaphragm is punctured.
- Diaphragm has become inflexible (due to deposits of solids).
- Movement of actuator is short/slow.
- Problem with hydraulic supply pressure.

Peristaltic Pumps

A peristaltic pump is a type of positive displacement pump used for pumping a variety of fluids wherein the fluid being transported does not come into direct contact with any part of the pumping mechanism. Hence it can be used for pumping sterile/pure fluids for medical and laboratory applications.

It consists of a flexible tube fitted inside a circular pump casing in which a rotor with a number of rollers attached to its circumference keeps rotating. The rollers compress the flexible tube as they rotate in the pump casing. Generally, these pumps have at least two rollers. They can have more rollers as per design for compressing (and releasing) the tube and in the process trapping some fluid between them.

The part of the tube under compression gets squeezed thereby forcing the fluid to move through the tube. As the roller rotates it moves away from the tube and the tube restores its shape again. This draws in more fluid into the tube. Compression by the next roller moves the fluid through the tube towards the pump exit nozzle (Fig. 5.8).

Pumps with Soft Tube

Peristaltic pumps for lower pressure typically have dry casings and non-reinforced tubes. The rollers attached to the rotor compress the tube when they move over the tubes during rotation of the rotor.

Pumps with Stronger Tubes

These have hoses (reinforced tubes) instead of tubes and hence can operate against considerably higher pressures. They have rollers in the shape of shoes.

Good quality lubricant with proper viscosity is filled in the casing. It is on the exterior of the hose and can prevent abrasion of the hose and helps to dissipate the heat caused by repeated compression by the rollers.

Chemical Compatibility
The pumped fluid contacts only the inside surface of the tubing. The fluid does not come in contact with pump parts like valves and seals. Therefore, it is necessary to check only the chemical compatibility of material of construction of the tubing with the fluid. This can prevent contamination of the fluids by the material of the tube.

The tubing should be able to maintain its circular cross-section even after repeated squeezing and restoration of shape (flexing) during operation of the pump. Hence materials like PTFE, PVDF *cannot* be the ideal choice for the tubes even though they are compatible with many chemicals.

Materials such as hypalon, natural rubber, and polypropylene may be considered for use as tube materials. Natural rubber has good resistance to repeated flexing while hypalon has good chemical compatibility. However, tubes of materials like viton cannot withstand the repeated flexing operations in the pumps which run continuously. These may not be selected on the basis of having good compatibility with acids and hydrocarbons alone.

Hence better types of lined tubes having resistance to many chemicals and which can also withstand the repeated flexing operations are to be provided. These tubes shall have thin inner layers of PTFE and a stronger outer layer. However, such lined tubes are generally costly and the additional cost shall be considered against the longer life they offer.

Limitations
Sharp abrasive particles in the fluid being pumped can damage the inner layer and the outer layer can come in contact with the fluid through the leak.

Certain inner lining material can also crack or get detached from the outer layer due to repeated compression by the rollers. This can create contact of chemicals being pumped with the outer layer and contamination can occur.

Hence a sample of the tubing should be tested by immersing in the fluid (which is to be pumped) up to 48 h at the operating temperature. The loss in weight should be checked at end of the test. If there is a loss of more than 3–4% the tube is not to be used with the fluid to be pumped.

Flow Rate
The flow rate of a peristaltic pump is higher when:

- Inner diameter of tube is larger
- Diameter of nozzles is larger
- Higher speed of rotor

Applications
Peristaltic pumps are typically used to pump clean/sterile or highly reactive fluids as these do not come in contact with pump components like valves and "O" rings. This prevents contamination of the fluids being pumped. This enables use for pumping reactive chemicals, slurries. They are also used in medical applications for blood circulation.

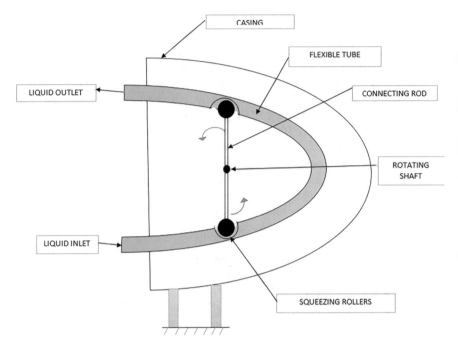

Fig. 5.8 Peristaltic pump

Caution
The tube can wear out slowly due to repeated compression and flexing back to original shape due to movement of the rollers. This can diminish the feed rate when the tube deteriorates. Regular replacement of the tubes may be required specially when slurries with abrasive particles are pumped.

Progressive Cavity Pumps
Most types of positive displacement pumps have close tolerances and an all metal construction. Hence, they cannot handle viscous slurries with suspended (and abrasive) solids. Gear pump and centrifugal pumps can wear out when such solids are present in the liquid. The gear pumps can also get clogged. Gear pumps are provided with internal pressure relief valve which is generally spring-loaded type or a rupture disc in the discharge line. A pressure switch is provided in the discharge line to trip (switch off) the drive motor in case of excessive pressure caused by clogging of discharge line, delivery valve closed by mistake, etc.

Centrifugal pumps are not able to efficiently handle highly viscous fluids. The discharge flow rate gets reduced while they consume more power when the viscosity is high.

Progressive cavity pumps are a type of positive displacement pump which are designed to handle highly viscous fluids. They generally consist of a helical rotor and a stator. The rotor moves in close contact tightly against the stator as it rotates, forming a set of cavities which move when the rotor is rotated.

The rotor of the pump is normally made of steel and is provided with a smooth hard surface. The stator body of the pump is made of a moulded elastomer/rubber inside a metal tube body.

Main Characteristics

The moving progressive cavities deliver a uniform and non-pulsating flow. The developed head is independent of the rotational speed, whereas the capacity is proportional to the speed. It is therefore possible to control the discharge flow by providing variable frequency drives (VFDs) for adjusting the pump speed.

Non-clogging

These pumps can handle highly viscous slurries if the fluid component can almost enclose (cover) the abrasive particles and the concentration of such solids in the medium is not large. *However, abrasive particles in large concentration can shorten the life of the stator.*

Hence provide a sample of the slurry to the pump manufacturer and request for a suitable model.

Low Internal Velocity

Shear-sensitive materials can be handled with minimum degradation due to low internal velocity in the pump. This is because the medium flows axially and is not moving in the casing at a high speed. The abrasive particles move at low speeds in the pump, and generally do not erode the internals.

Less Noise During Running

The rotors turn inside stators which are made of resilient material (e.g. rubber) and hence do not generate much noise during operation.

These pumps require lower net positive suction head (NPSH) as compared to a centrifugal pump. Consult the manufacturer while procuring such pumps and request for appropriate model of the pump which will be able to handle the suction lift in the given location.

Precautions

These pumps can push through the viscous medium at a high pressure. Hence, they should be fitted with protective devices like high discharge pressure trip switches for the motor, rupture discs, or a bypass line which can return the medium to the inlet side.

Make sure that the pump will never run dry, as it can result in deterioration of the stator.

Typical Applications

- Wastewater treatment plants: For pumping of sewage and sludge.
- Pumping of viscous chemicals.
- Pumping of slurries from coal mines, oil wells, etc.

5.15.2.2 Rotodynamic Pump

Centrifugal Pump

This is a common example of rotodynamic or dynamic pressure pump. Centrifugal pump converts energy from an electric motor first into velocity or kinetic energy and then into pressure energy of fluid that is being pumped.

The energy change occurs by virtue of two main parts of pump—the impeller that converts driver energy into the kinetic energy and volute or diffuser which is the stationary part that converts kinetic energy into pressure energy. The basic principle on which centrifugal pump works is that the liquid is made to rotate by an external force, then thrown away from the central axis of rotation and a centrifugal head is imparted which raises the liquid to a higher level.

In addition to the centrifugal action, as the liquid passes through the revolving member or impeller, its angular motion changes which also results in increase in pressure of the liquid.

Energy added by the centrifugal force is converted to kinetic energy. The amount of energy given to the liquid is proportional to the velocity at the edge or vane tip of the impeller. Faster the impeller revolves or the bigger the impeller is, then higher will be the liquid velocity at the vane tip and greater energy is imparted to the liquid. This kinetic energy of liquid coming out of impeller is harnessed by creating a resistance to the flow. The first resistance is created by the pump volute (casing) that catches the liquid and slows it down. In the discharge nozzle, the liquid further decelerates, and its velocity is converted to pressure energy according to Bernoulli's principle. Therefore, the head (pressure in terms of height of liquid) developed is approximately equal to the velocity energy at the periphery of the impeller expressed as:

$$H = V^2 / 2g$$

where, H = total head developed; V = velocity at periphery of impeller; g = acceleration due to gravity.

Main Components of Centrifugal Pumps

1. Rotating components consisting of impeller, shaft, and shaft coupling.

   - Impeller: This is the main rotating part that provides the centrifugal acceleration to the fluid. Impeller is a wheel provided with blades or vanes and is mounted on a shaft which is coupled to an external source of energy.
   Impeller classification:
   Based on major direction of flow in reference to the axis of rotation

   – Radial flow
   – Axial flow
   – Mixed flow

   Based on suction type

   – Single-suction: Liquid inlet on one side.

– Double-suction: Liquid inlet to the impeller symmetrically from both sides. In this the end thrust on shaft is reduced and also inlet velocities are reduced as the net inlet flow area is larger.

Based on mechanical construction

– Closed: Shrouds or sidewall enclosing the vanes. (for clean liquids)
– Open: No shrouds or wall to enclose the vanes (for liquids with suspended solids)
– Semi-open or vortex type (for dirty liquids with suspended solids)

  • Shaft: The shaft transmits torques encountered when starting and during operation while supporting the impeller. It must do this job with a deflection less than the minimum clearance between the rotating and stationary parts.

Multistage centrifugal pump has two or more identical impellers mounted on the same shaft or on different shafts.

Coupling: This connects centrifugal pump to pump driver. Shaft couplings can be rigid or flexible.

2. Stationary components consisting of casing, casing cover, and bearings.

  • Casing: This is a straight chamber, which surrounds the impeller. The gradually increasing cross sectional area of casing helps in maintaining uniform velocity of the flow. Casings are generally of two types: circular type and volute type. Volute casings are used to build a higher head and circular casings are used for low head. Inside the casing, circular casing has stationary diffusion vanes surrounding the impeller periphery that convert velocity energy to pressure energy.
  • Seal chamber and stuffing box: Seal chamber and stuffing box form a chamber, either integral with or separate from the pump case housing that forms the region between the shaft and casing where sealing media are installed. The seal chamber and stuffing box protects the pump against leakage at the point where the shaft passes out through the pump pressure casing. In seal chamber, the sealing is achieved by means of a mechanical seal and in a stuffing box the sealing is achieved by means of packing. When the pressure at the bottom of the chamber is below atmospheric, it prevents air leakage into the pump, and when the pressure is above atmospheric, the chamber prevents liquid leakage out of the pump.
  • Bearings keep the shaft or rotor in correct alignment with the stationary parts under the action of radial and transverse loads.

Single and Multistage Pump Single-stage pump has only one impeller mounted on a shaft and multistage pump has more than one impeller mounted on same shaft. The stages operate in series i.e. discharge from first stage flows into suction of second stage and so on.

Series arrangement: In a multistage pump the impellers are mounted on the same shaft and enclosed in the same casing thereby producing heads greater than that available with a single impeller but the discharge remains constant. The heads by individual impellers add up. The multistage pump is used to deliver small quantity of liquid with high head.

e.g. High pressure boiler feed water pumps.

Note: These high pressure multistage centrifugal pumps (typically for boiler feed water) generally have a small recycle line with a valve to recycle the liquid back to suction side (going back to the feed water tank) if the discharge valve is closed/ discharge side is blocked due to some reason. This is to have a certain minimum flow of liquid to prevent overheating of the pump.

- **Parallel arrangement**: More than one pumps are run to discharge a large quantity of liquid against same head. This is used for pumping large quantity of liquid with smaller head. The flow rates may be different and are used if varying delivery is to be made.

Note: Non-return valves and stop valves shall be provided in the delivery side of each pump.

Operating Characteristics of Centrifugal Pump

Characteristic curves of centrifugal pumps predict the behaviour of performance of pump under different operating conditions. The overall efficiency of pump is due to—hydraulic efficiency, mechanical efficiency, and volumetric efficiency.

Mechanical efficiency is based on losses in bearings and seal due to which power by the impeller is less than the power produced by the driver i.e. absorbed by the pump shaft.

Hydraulic losses occur due to loss of energy of fluid due to friction between fluid and the walls, and change of fluid flow direction. Volumetric losses occur due to leakage of fluid between impeller and the casing and other pump components and thus all the flow that enters the pump does not exit from it.

- In order to obtain the main characteristic curve of a pump, it is operated at different speeds. For each speed, the rate of flow (Q) is varied by means of delivery valve and for different values of Q, the corresponding values of discharge head (H_m), shaft horse power (P) and overall efficiency (η_o) are measured or calculated. The same operation is repeated for different speeds.
- Then H_m v/s Q, P v/s Q and η_o v/s Q curves for different speeds are plotted which represent the main characteristics of pump and indicate the performance of the pump at different speeds (Fig. 5.9).
- During operation, the pump is generally required to run at a constant speed which is its designed speeds. The speed is varied as per need by a variable frequency drives (VFD) in certain installations.

Advantages of Centrifugal Pump

- Simple in construction.
- No valve action in operation.
- Maintenance is easy.
- Can handle liquid with suspended solids.
- Discharge is uniform without shocks or pulsations.

– This pump is cheaper.

Disadvantages of Centrifugal Pump
– Not useful for highly viscous liquid.
– Maximum efficiency over a narrow range of condition.
– Priming is required.

Selecting a Pump Impeller
When selecting the size, never select a pump that requires maximum size of impeller, which can make future modification to the pump impossible to increase the output further.

Selecting a Pump Motor
- Select a motor for centrifugal pump based not on the size of impeller used but on the maximum impeller diameter that will fit into the pump.
- The reason for selecting higher KW rating motor are:

  – It does not trip for higher initial torque.
  – There is provision for further expansion of the pump capacity.

General Provide variable frequency drives for the motor which can control the speed *as per required output from the pump. This method can control the power consumption.* Check if the cost of VFD can be recovered within a reasonable time (say about 1 or 2 years)

Shut Off Pressure
Shut off pressure is the maximum pressure that the pump will develop at zero flow at the outlet (when the delivery valve is fully closed) and is the critical variable to be considered when selecting the size of an impeller. It represents the rise in pressure head across the pump with discharge valve closed i.e. at zero discharge. Larger the impeller, larger the shut off pressure.

- Shut off pressure is taken as the design pressure (usually set at the range of 1.1 to 1.3 times the maximum operating pressure of the equipment) at discharge line as it is the maximum pressure that the pump can develop.
- For the illustration of importance of shut off pressure, consider Fig. 5.10

If operator accidentally closes the valve at outlet of the heat exchanger, the shell of heat exchanger shall be subjected to shut off pressure of the pump. In case heat exchanger shell is not designed for shut off pressure, we get two options:

– Use small impeller
– Install relief valve on shell

A design engineer cannot ignore the consequences on downstream equipment when expanding the capacity of the pump. The need to design all process equipment (*capable to withstand the pump shut off pressure*) between pump and a block valve at outlet of the process equipment is a necessary requirement for protection of the equipment.

Expanding Pump Capacity

There are two methods to increase pump output:

- Reduce downstream pressure drop: The typical causes of excessive pressure drop and suggested remedies are listed below:

  - High tube side pressure drop through a shell and tube heat exchanger: Reduce number of tube side passes if a new heat exchanger can be designed and provided.
  - High shell side pressure drop in a shell and tube heat exchanger: A new tube bundle with increased baffle spacings if a new heat exchanger can be designed and provided.
  - High pressure drop through a wide open control valve: Change control valve port size or trim to the maximum size permitted in the control valve body.
  - Reduce excessive piping losses: Minimise number of fittings, sharp bends in piping, and entry losses in equipment.
  - Increase pipe diameter or provide parallel piping runs. This can be expensive.

- Increase size of impeller:

  - Check the effect of larger impeller when expanding a pump capacity with capacity of existing drive motor. Changing of impeller is easy and cheaper when compared to changing the motor and its associated components, e.g. starter, cables, and base plate, as the total cost can be more.
  - Increasing the diameter of impeller can result in higher flow rate, discharge pressure, and power requirement as given by following empirical relations.

    $$Q_2 = Q_1 \left(d_2 / d_1 \right)$$

Fig. 5.9 Centrifugal pump characteristics

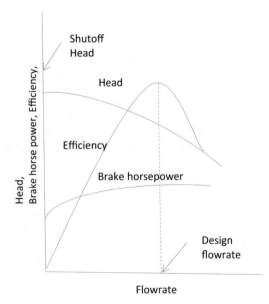

$$h_2 = h_1 \left(d_2 / d_1 \right)^2$$
$$P_2 = P_1 \left(d_2 / d_1 \right)^3$$

where: Q_2 = Flow rate with larger impeller; Q_1 = Flow rate with existing impeller; d_2 = Diameter of larger impeller; d_1 = Diameter of existing impeller; h_2 = Head delivered by larger impeller; h_1 = Head delivered by existing impeller; P_2 = Power drawn by motor with larger impeller; P_1 = Power drawn by motor with existing impeller.

Increase speed of the pump in consultation with the pump manufacturer. This may need a motor with higher rating. This option needs careful study.

Cavitation

Cavitation is the formation of bubbles in a liquid and their subsequent collapse in the pump. These bubbles move into regions of higher pressure towards pump discharge where they collapse suddenly causing damage to moving parts of the pump.

There are two types of cavitation:

- Vaporous cavitation: Vapor bubbles are formed due to the vaporisation of the process liquid which is being pumped. It occurs when the pressure on the liquid entering the eye of centrifugal pump gets reduced and falls below the vapor pressure of the liquid causing the liquid to flash and form bubbles. This cavitation condition induced by formation and collapse of vapor bubbles is referred to as vaporous cavitation. It results in reduced unsatisfactory pump performance, excessive noise and vibrations and wear of pump parts.
- Gaseous cavitation: Gas bubbles are formed due to the presence of dissolved gases (air or any gas) in the liquid which is being pumped. The cavitation condition induced by the formation and collapse of gas bubbles is commonly referred to as gaseous cavitation. Gaseous cavitation can result in loss of capacity.

Priming

Priming is removing air from the suction line and pump casing. If the impeller is running with air, then it will produce negligible pressure at the suction and cannot suck liquid. This can be avoided by priming. This is done by first filling the suction pipe and casing with liquid being pumped. Then the pump is started with the delivery valve almost closed (only slightly open) so that the rotating impeller pushes the liquid with all the air, gas, or vapor contained in the passageways of pump in the delivery pipe. Finally, delivery valve is opened further so that air is displaced by the liquid to be pumped and escapes out.

The pump would not function properly when not completely filled with liquid and if the pump is allowed to run without fluid, it will overheat the pump system and there will be a danger of damage to critical internal pump components.

Vertical submerged centrifugal pumps do not need priming since they are generally submerged in the liquid in the tank. However, the level of liquid in the tank must be maintained as per minimum submergence level recommended by the manufacturer.

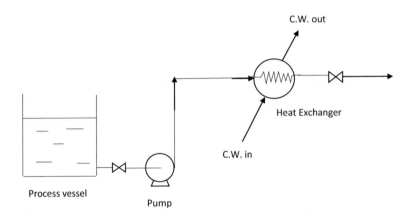

Fig. 5.10 Unintentional shut off pressure/accidental creation of shut off pressure

Pumps Used for Hot Liquids

The pump and motor are fixed on the base plate at ambient temperature during erection. However, the pump can become hot during operation and the alignment with the motor can get disturbed.

Hence the pump is placed slightly lower than motor during the initial stage. The pump will expand slightly as it gets warmed up and further after running. After that the shafts of the pump and the motor will come in line with each other.

The initial misalignment is carried out by keeping thin metallic shims below the motor. Information about the amount by which the motor is to be slightly raised up is generally provided by the original equipment manufacturer.

In case the OEM has not provided this information a dial gauge can be mounted on the pump shaft in vertical, sideways, and in axial direction. The pump warm-up is allowed to take place after this. The movement of dial gauge indicates where and how much the pump is moving. The shims below the motor are adjusted accordingly.

Later on, as a good maintenance practice, the pump is run for a few hours or days and alignment re-checked because now the pump has achieved a steady temperature.

Problems with Centrifugal Pumps

Pump does not deliver or delivers too less
  Typical reasons

- Too high suction lift.
- Pump not primed properly (not filled up fully with liquid at start).
- Air pocket in suction line (suction line has a high point before it meets pump suction opening) which obstructs liquid flow.
- Liquid has become too viscous due to low ambient temperature.
- Specific gravity of liquid larger than the designed value.
- The viscosity of the liquid is larger than the designed value.
- Corroded/damaged impeller.

- Increased clearance between impeller and casing wear rings.
- Suspended particles have choked the impeller vanes.
- Malfunctioning of foot valve in suction side.
- Strainer in suction line is clogged.
- Clogged sealing liquid line for shaft seal.
- Suction line is clogged/has too small diameter.
- Air ingress through leaking suction line.
- Vapour lock in suction line due to high temperature of liquid.
- Clogged discharge line.
- Too high discharge head (more than design head of the pump).
- Discharge valve remaining fully/almost closed due to any reason.
- Malfunctioning of valves in suction and/or discharge side.
- Discharge valve of standby pump is partially open.
- Coupling with motor is damaged.
- Wrong direction of rotation of motor (used for driving the pump).
- Drive motor of low power rating (lesser HP than required).
- Drive motor has lower speed than pump design speed.

Too many vibrations
Typical reasons

- Air ingress through leaking joints in suction line.
- Air ingress through leaking shaft seal.
- Cavitation
- Pump being operated below minimum continuous stable flow.
- Unbalanced rotor.
- Damaged impeller.
- Bearing failure of motor or pump.
- Piping inadequately supported/tightly fixed.
- Loose foundation bolts or base plate
- Improper alignment with drive motor.
- Operating at too low discharge pressure.
- Worn or loose bearings.
- Bent shaft.

Drive motor is overloaded/tripping frequently
Typical reasons

- Higher specific gravity than design.
- Higher differential head.
- Low pump efficiency due to wear of internals.
- Higher discharge flow.
- Mechanical damage to driver or pump.
- Clogged impeller.
- Too viscous liquid (*liquid can become very viscous due to low ambient temperature/due to non-functioning of steam traps on liquid lines in winter*).
- Pump being run at higher speed than recommended by manufacturer.

- Very low discharge line pressure.
- Clogged suction Line.
- Closed/throttled valve in discharge line.
- Over-tightened gland packings.
- Trip relay set at lower current.
- Low supply voltage (causes the motor to draw high current).
- Drive motor of lower HP provided.

Too much fluctuations in discharge flow
Typical reasons

- Clogged suction or discharge lines.
- Malfunctioning of valves in suction or discharge side.
- Malfunctioning of flap in foot valve.
- Fluctuating low level in liquid tank at inlet side with occasional air ingress in suction line of the pump.
- Loose coupling connection with motor.
- Foreign material in the pump casing/impeller.
- Improper alignment of the pump and motor.
- Unbalanced rotor assembly.
- Damaged impeller
- Loose gland packing allowing air ingress.

Pump discharge pressure low

- Too low a speed
- Worn wearing rings
- Damaged impeller
- Worn packing
- Gas or vapor in the liquid.
- Too viscous a liquid.
- Wrong direction of rotation.
- Impeller diameter too small.
- Obstruction in water passages.
- Worn gaskets.

Pump loses prime after starting
Typical reasons

- Incomplete priming.
- Too high suction lift.
- Air leaks in suction pipe or packing glands and gas or air in liquid.
- Inlet not sufficiently submerged.
- Low available NPSH.
- Plugged seal liquid piping.

Stuffing box getting overheated
Typical reasons

- Too tight packing.
- Not enough packing lubricant.
- Wrong grade of packing.
- Insufficient seal liquid flowing to the packing.
- Incorrect installation of packing.
- Bearings overheat.
- Too low oil level.
- Poor or wrong grade of oil, dirt in the bearings or the oil, moisture in the oil
- Failure of the oiling system.
- Not enough bearing cooling water.
- Bearings too tight, misalignment, or oil seals fitted too closely on the shaft.

Practical Considerations for Installation of Centrifugal Pumps
- The foundation must be level and fully cured.
- The foundation bolts must be of sufficient length and diameter as advised by the supplier of the pump. They shall be firmly grouted in the foundation for providing the base plate which will be used for installing the pump and the drive motor.
- The drive motor and pump must be accurately aligned to prevent vibrations, bearing failure and other mechanical damage. The alignment shall be checked by dial gauge.
- The suction line should have a diameter at least twice that of discharge nozzle of the pump. Velocity of the liquid in the suction line should not exceed 2 m/s.
- The bend in suction line should be long radius type. Sharp 90° bend should not be used.
- The bend in suction line should be long radius type. Sharp 90° bend should not be used.
- The suction line should be well supported and have minimum joints. It should be fabricated only after the pump has been positioned and fixed on the base plate. This is to avoid causing stresses on the pump.
- A proper foot valve should be provided in case of a suction lift (negative suction head). Either a priming vessel (with sufficient volume) or an auxiliary source of liquid (water) should be available to fill up the suction line and the pump before starting.
- The discharge line should be well supported so that it does not cause any stresses on the pump. It should have a non-return (check) valve fitted to prevent back flow of liquid (Figs. 5.11 and 5.12).

Installation of Vertical Submerged Centrifugal Pumps
These are typically mounted on process tanks, wells and storages, etc. The pump is submerged in the liquid to be pumped and hence does not require priming. However, they can fail to deliver liquid if the level falls below the minimum submergence level recommended by the manufacturer.

A strainer shall be provided for the suction nozzle. It should preferably have a screen of corrosion resistant material.

Base plate of these pumps must be independently supported by providing external support structure outside the tank (especially when it is a brick lined tank). This is to prevent loosening of the brick lining due to vibrations caused by running of the pump.

An external gantry structure with a horizontal beam shall be provided outside the tank for taking out the pump for maintenance (if required). A travelling chain pulley block should be available on the beam (Fig. 5.13).

Problems with Vertical Submerged Pumps

Pump does not deliver or delivers too less
    Typical reasons

- Pump is not submerged properly in the liquid.
- Liquid has become too viscous due to low ambient temperature(winter).
- Specific gravity of liquid more than the designed value.
- The viscosity of the liquid is more than the designed value.
- Corroded/damaged impeller.
- Increased clearance between impeller and wearing rings.
- More level of thick sludge in the tank.
- Suspended particles have choked the impeller vanes.
- Suction nozzle strainer is clogged.
- Suction nozzle has too small diameter.
- Clogged discharge line.
- Too high discharge head (more than design head of the pump).
- Discharge valve remaining fully/almost closed due to any reason.
- Malfunctioning of valves in discharge piping.
- Coupling with motor is damaged.
- Wrong direction of rotation of motor (used for driving the pump).
- Drive motor of low power rating (lesser HP than required).
- Drive motor has lower speed than pump design speed.

Drive motor is overloaded/tripping frequently
    Typical reasons

- Higher specific gravity of liquid than design.
- More level of thick sludge in the tank.
- Wear of internals.
- Higher discharge flow.
- Mechanical damage to driver or pump.
- Clogged impeller
- Too viscous liquid (*liquid can become very viscous due to low ambient temperature/due to non-functioning of steam traps on liquid lines in winter*).
- Pump being run at higher speed than recommended by manufacturer.
- Very low discharge line pressure.
- Clogged suction line.
- Trip relay set at lower current.
- Drive motor of lower HP provided.
- Low supply voltage (causes the motor to draw high current).

Too many vibrations
Typical reasons

- Oblique shaft.
- Damaged impeller.
- Bearing failure of motor or pump.
- Piping inadequately supported/not properly fixed.
- Loose base plate (not properly supported.
- Improper alignment with drive motor.
- Operating at too low discharge pressure.

5.15.3 Different Types of Heads in Pumping Calculations

Head is a measurement of the height of a liquid column that the pump could create from the kinetic energy imparted to the liquid.

Suction static head (h_S): Head resulting from elevation of the liquid relative to the pump center line on the pump suction side. If the liquid level is above pump center-line, h_s is positive. If the liquid level is below pump centerline, h_s is negative. Negative h_s condition is commonly denoted as a "suction lift" condition.

Discharge static head (h_d): This is the vertical distance between point of free discharge or the surface of the liquid in the discharge tank relative to the pump center line.

Friction head (h_{fs}): This is the head required to overcome the frictional resistance to moving fluid in the pipe and fittings. It is dependent upon the size, condition, and type of pipe, number and type of pipe fittings, flow rate, and nature (viscosity) of the liquid.

Vapor pressure head (h_{vp}): Vapor pressure is the pressure at which a liquid and its vapor co-exist in equilibrium at a given temperature. When the vapor pressure is converted to head, it is referred to as vapor pressure head. The value of h_{vp} increases with temperature and in effect, opposes the pressure on the liquid surface, the posi-tive force that tends to cause liquid flow into the pump suction *i.e. it reduces the available suction pressure head.*

Pressure head (h_{ps}): Pressure head must be considered when a pumping system either begins or terminates in a tank which is under some pressure other than atmo-spheric. The pressure in such a tank must first be converted to meters of liquid. Pressure head refers to absolute pressure on the surface of the liquid reservoir sup-plying the pump suction, converted to meters of head. If the system is open, h_{ps} equals atmospheric pressure head.

Velocity head (h_{vs}): Refers to the energy of a liquid as a result of its motion at some velocity. It is the equivalent head in meters through which the liquid would have to fall to acquire the same velocity, or in other words, the head necessary to accelerate the liquid. The velocity head is usually insignificant and can be ignored in most high head systems. However, it can be a significant factor and must be

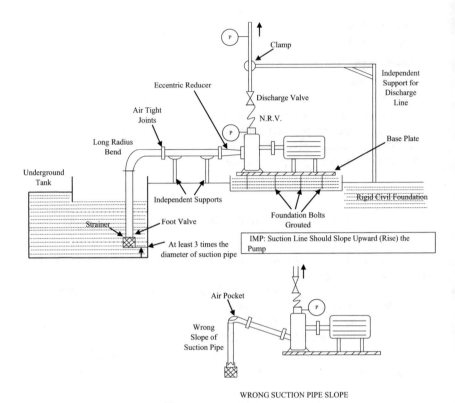

Fig. 5.11 Installation of centrifugal pump with suction lift

considered in low head systems. Velocity at which the pipe discharges into the over-head tank should be minimised.

Total suction head (H_S): This the sum of suction reservoir pressure head (hp_S), suction static head (h_S), and velocity head at the pump suction flange (h_{vs}) minus the friction head in the suction line (h_{fS}).

This is given by the following relation,

$$\mathbf{H_S = h_{pS} + h_S + h_{vS} - h_{fS}}$$

H_S = Total suction head; h_{pS} = Suction reservoir pressure head; h_S = Static suction head; h_{vS} = Velocity head at the pump suction; h_{fS} = Friction head in the suction line.

Net Positive Suction Head (NPSH)

NPSH allows a prediction to be made regarding the safety margin required to avoid the effects of cavitation during operation. Cavitation occurs due to formation of vapor bubbles when the liquid pressure at a given location is reduced to the vapour pressure of the liquid and causes noise, impeller damage, and impaired pump

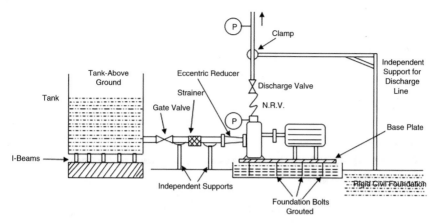

INSTALLATION OF HORIZONTAL CENTRIFUGAL PUMP WITH POSITIVE SUCTION HEAD AVAILABLE

Fig. 5.12 Installation of centrifugal pump with positive suction head available

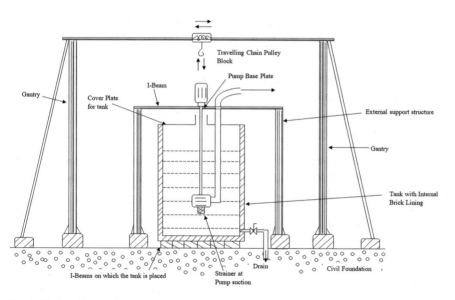

Fig. 5.13 Installation of vertical submerged pump

performance. For proper pump performance cavitation should be avoided by operating pump so that NPSH must be maintained or exceeded.

NPSH is difference between total head (sum of pressure head at the pump inlet, static head, and velocity head) at suction side and the vapor pressure head and friction head in suction line.

$$\mathbf{NPSH} = \mathbf{h}_{pS} + \mathbf{h}_S + \mathbf{h}_{vS} - \mathbf{h}_{fS} - \mathbf{h}_{vp}$$

Refer Appendix "Sample Calculation 2" for Sample problem on Pump power calculations

5.16 Fans and Blowers

Introduction
Various types of blowers and compressors are used in chemical industries for handling air and various gases.

Typical uses are for providing air for combustion of fuels (Combustion Air fans), for carrying out reactions (in process plants for manufacturing sulphuric and nitric acids), for operating waste destruction plants (as induced draft fan); for wastewater treatment plants (for aeration), for supply of compressed air to pneumatically operated valves, etc.

5.16.1 Guidelines for Installation/Erection and Commissioning

- Check purchase order and name plate on the machine. Confirm that the correct machine is procured.
- Discharge capacity, operating pressure, temperature, composition of gases, operating speed to be confirmed.
- In particular check if the gases contain corrosive components and moisture; and possibility of condensation of acidic particles at operating temperature.
- Confirm that civil foundation is properly cured and clearance is given by civil engineer to place load on it.
- Confirm that positions of foundation bolts are as per installation drawing. These bolts must be properly grouted. Base plate of blower must be securely fitted on the foundation.
- Orientation of suction and discharge nozzles to be checked. Provide independent supports for suction and discharge ducts so that there is no load on the blower.
- Confirm that the material of construction of casing, impeller, shaft, etc. are suitable for gases to be handled.
- Typical options are carbon steel, stainless steel, rubber lined, fiberglass reinforced plastic impellers, etc.
- Tip speed of the rubber lined impellers must be discussed with the manufacturer because the lining can come off at high tip speeds. Confirm the rating (HP) and speed of drive motor.
- Check the levelling as well as alignment with the blower. The coupling/belt drive arrangements shall be checked.

- Variable frequency drive controller may be provided for speed control of the blower to obtain required discharge capacity while trying to minimise power consumption.
- Provide filter on suction side (if dust particles are present in the gases) and pulsation dampener/surge vessel on discharge side if recommended by manufacturer.
- Flow control and non-return valves shall be provided on discharge side for parallel operation of blowers.
- A centrifugal blower can overload the drive motor if the discharge pressure is too low. Hence provide a control valve, a pressure gauge (and a flow meter), on discharge side to monitor the flow and an ammeter to check electrical load on the drive motor.
- It is advisable not to start a centrifugal blower with the discharge valve in full open position specially in case of very low discharge pressure. The discharge valve shall be almost closed at start and opened slowly thereafter.

5.16.2   Positive Displacement Type Blowers

General checks for installation are similar to checks for centrifugal blowers.

These should be provided with safety valve and an electrical trip arrangement as a precaution against high discharge pressure.

A pulsation dampener and surge vessel shall be provided in consultation with piping system designer. Flexible distance piece of rubber is provided in the discharge line.

Gases released from the safety valve shall be vented through a suitable scrubber away from the working area. This precaution is necessary while handling toxic/dangerous gases.

An inter-stage cooling arrangement shall be provided to control the temperature of the gases when a multistage compressor is used.

5.16.3   Reciprocating Compressor

Pulsation damper should be provided by original equipment manufacturer (OEM).

Independent supporting structures should be provided for the piping.

5.16.4   Screw Compressor

These also have pulsation dampener as they are positive displacement machines. This should be provided by OEM.

A screw compressor can be used when smooth pulsation free flow is required and the discharge pressure is not too high.

However, an inter-stage gas cooler cannot be provided for a single-screw compressor.

The inter-stage cooler can be provided when more than one-screw compressors are operated in series.

Notes Twin lobe blowers and three lobe blowers are examples of positive displacement type blowers. The discharge from three lobe blowers has less pulsations than discharge from twin lobe blowers; and also makes less noise.

Positive displacement type blowers should never be started when the discharge valve is closed.

References/for Further Reading

1. "Process Equipment Procurement in the Chemical and Related Industries", Kiran Golwalkar, Published by Springer International Publishing, Switzerland.
2. "Production Management of Chemical Industries", Kiran Golwalkar, Published by Springer International Publishing, Switzerland.
3. Coulson and Richardson's Chemical Engineering Design,R.K.Sinnott Volume 6, Fourth Edition Vol.6 , Butterworth- Heinemann.

Chapter 6
Cooling and Heating Systems

Abstract Cooling systems are required in chemical plants to control temperatures of process units, for condensation of volatile materials, for distillation operations, etc. Various facilities such as spray ponds, cooling towers, refrigeration plants are used for cooling various process streams and materials. These need careful observation during operation and proper precautions before taking up maintenance work. Unconventional units based on vapour absorption chilling are also used in plants where waste steam is available. These operate on the principle of cooling effect produced by boiling water under vacuum. The water vapours are absorbed by strong LiBr solution to maintain the vacuum. The dilute LiBr solution is concentrated by steam heating.

Chemical plants need heating arrangements for carrying out reactions, drying and melting of raw materials, evaporation of dilute solutions, etc. These operations are to be carried out at different temperatures and may need accurate control of the temperatures. Cost of heating depends on calorific value and infrastructure required for storing and handling the heating agents.

Facilities for land and air pollution control are also required depending on the fuels.

Keywords Cooling systems · Spray ponds · Cooling towers · Maintenance of cooling towers · Conventional refrigeration plants: precautions for maintenance work · Vapour absorption chiller: check points for proper working · Heating systems · Design criteria · Electrical heating · Steam heating · Hot thermic fluids · Coal fired systems · Oil fired systems · Gas fired System · Heating by hot process gases · Pollution control required · Storage and handling arrangements

6.1 Introduction

The plant engineers should estimate the total cooling and heating load and also consider the required temperature of the process streams before selecting the cooling and heating system. Any variation as per seasonal climatic conditions and production rate of the plant shall also be considered. These operations are to be carried out at different temperatures and may need accurate control of the temperatures.

Cooling Systems
These are required in many chemical industries for:

- Carrying out certain reactions at very low temperatures.
- Condensation of volatile products.
- Recovery of valuable chemicals from exit gases of process units.
- Absorption towers which need cooled circulating liquids for better absorption.
- Cooling of storage tanks containing volatile chemicals.

Heating Systems
Chemical plants need heating arrangements for

- Carrying out reactions.
- Drying and melting of raw materials.
- Evaporation of dilute solutions, etc.

Cost of heating depends on calorific value and infrastructure required for storing and handling the heating agents.

6.2 Practical Considerations for Cooling Systems

- Cooling load expected at ambient temperature: Typically, this type of cooling load is to be expected for process streams and equipment which are to be cooled/maintained at temperatures of 25–40 °C. These include cooling of reactor vessel, cooling of storage tanks by water spray, condensation of vapours around 50 °C, and cooling of electrode glands. Such cooling can be carried out by circulation of cold water from a cooling tower/cooling spray pond.
- Cooling requirements at low temperatures (between 5–20 °C) are generally met by chilled water which is produced by refrigeration.
- Cooling requirements below 5 °C are generally handled by brines and freezing mixtures.
- Initial cooling should be done by exchange of heat between hot outgoing process stream and incoming cold process stream as far as possible.
- *Example: Further cooling of hot condensed refined CS_2 from a refining unit can be done by cold incoming CS_2 to the refining unit.*
- This initial cooling shall be followed by further cooling which can be done by circulating cold water from cooling tower.

- Subsequently, additional cooling can be achieved by chilled water or special brines.
- Cooling systems consisting of spray ponds, cooling towers, and refrigeration units are therefore provided in chemical plants.
- Refrigeration systems are generally used for producing chilled water at (+5° to +10 °C) or brines at (−10° to −5 °C).

- *The energy consumption in refrigeration plants is more when operating at lower temperatures of the cooling media; hence select optimum range of temperatures.*

6.3 Spray Ponds for Cooling Water

- Warm incoming water from process units is sprayed through nozzles for cooling by evaporation and the cooled water is supplied to process units for cooling. However, this arrangement needs more space, and the water can get contaminated.
- The returning hot water from process units should not get mixed with the cooled water being supplied to the process units such as condensers or cooling jackets of reactors. Hence a suitable section of the spray pond is partitioned and designated as hot well. The returning hot water from process is admitted in this section. It is then sucked by pumps and sprayed through nozzles to cool it by evaporation. It now flows into another partitioned section designated as cold well. The cooled water is supplied to process units by another pump.
- Standby water pumps are provided for both hot well and cold well for smooth plant operations.
- Maintenance of spray pond system: Regularly remove the settled sludge and algae from a small settling chamber specially constructed in a corner of the pond. Spray nozzles shall be cleaned regularly to maintain satisfactory cooling of water. Make up water shall be provided by automatic control of level in the ponds. Add weedicides/algaecides to the spray pond water as per dose recommended by the supplier to prevent growth of algae.
- Advantage: Spray ponds provide a large storage of water in the plant for use in emergency.

6.4 Cooling Towers

These are used for cooling the hot water from exit of process units; from condensers in refrigerating system, etc. They are generally constructed of RCC basins with wooden structures on them; or of Fibreglass bodies with PVC fills. They are installed on ground floor or at a height (on the roof in some process plants).

6.4.1 Observations During Operation

- Keep a record of the temperature of the incoming warm water and cooled water at exit of the cooling tower during summer, winter, and rainy season. The effect on process plant (condensation of volatile liquids, cooling of reactor jackets, etc.) shall be monitored.
- Incoming warm water should be uniformly distributed over the fill material for efficient cooling. Check the distribution points.
- Regularly monitor the working of water pumps—discharge pressure, current drawn by drive motor, and any abnormal noise.
- (*Standby pumps should be operated alternately*)
- Water level and sludge level in basin.
- Analysis of incoming water which is used for make up to the cooling tower. Use only treated water for make-up. Ideally this should be from RO plants so that the heat transfer surfaces of process equipment remain clean. The higher cost of make-up water should be balanced against the advantage gained for better heat transfer (when the cooling water is used for condensation of valuable products).
- Suitable treatment shall be given to the circulating water in the cooling system to prevent scale formation and corrosion. Corrosion inhibitors such as chromates may also be added to prevent corrosion of process units being cooled by the water from the cooling tower. The dose of these chemicals shall be determined after analysis of the circulating water.
- Working of float operated control valve for make-up water.
- Analysis of the blow down water from cooling tower.
- Functioning of induced draft fan. Observe the air discharge from the ID fan.
- A cloth ribbon may be tied at top in the path of discharge air flow from the ID fan. Observe the fluttering of this ribbon tied at top. This will enable observation from a distance also. There should not be any recycling of the discharged air flow towards the air inlet sides of the cooling tower.
- Any vibrations in the ID fan, its gear box, and drive motor.
- Any vibrations in the body of the cooling tower.

Vibration sensors are to be provided for ID Fan—gear box—drive motor assemblies; and the body of cooling tower for monitoring the vibrations. Alarms shall be provided to warn the operating personnel if the vibrations exceed safe limits. It is advisable to keep a regular watch on the running of the ID Fan, lubricant level in gear box and current drawn by the drive motor.

- Any damage to the fill material and the support grid for fill.
- Any corrosion or weakening of supports of cooling tower when it is erected at a height or on the roof of the process plant.
- Analysis of water circulated in system (pH shall be neutral/alkaline; the total dissolved solids shall not generally exceed 150–200 ppm).

6.4.2 Maintenance of Cooling Towers

Alternate arrangement needs to be made for supplying cooling water to process units, refrigeration system, and cooling of storage tanks when the cooling tower is to be taken up for maintenance work. It may need isolation of the cooling tower completely OR isolation of the cells one by one if possible.

Water Pumps
- Check strainer in suction lines of the pumps.
- Check the foundation bolts, coupling halves.
- Any abnormal noise or vibrations.
- Provide fresh anti-corrosion coating on impeller and internals of the pump.
- Check the alignment with the drive motor.
- Check condition of the motor.

ID Fan
1. Check blade angle and adjust to optimize for reducing power consumption.
    Fibreglass reinforced plastic can also be used instead of metallic blades to reduce power consumption.
2. Install arrangement for automatically switching off the ID fan when the temperature of cooling water is low enough during winter.

Cooling Tower Basin
Do not allow buildup of sludge in the basin or high concentration of solids in the circulating water because these can deposit on heat transfer surfaces or can clog the tubes of process condensers. Remove the sludge from the lowest point provided and drain out some water from the cooling tower regularly. This water can be reused for floor washing or gardening in the premises.

Record of the Maintenance Work
Typically, it should include the following:

- Repairs to water circulation pumps and motors.
- Repairs to water pipe lines, valves, and pressure gauges on water lines.
- Repairs to ID fans, gear box and their alignment; bird cage on ID fan.
- Drive motor, its starter, cables and lighting.
- Repair to ladder for climbing up to check ID fan.
- Repairs to basin and tower fill materials.
- Temperature indicators and low temperature cut-out for ID fan.
- Cleaning of basin to remove accumulated sludge, and thorough washing thereafter.
- Float operated valve for make- up water.
- Hot water distributor (for hot water coming from process).
- Removal of any vegetative growth, algae, slimy material.
- Record of chemicals added for treatment of circulating water.

6.4.3   Fill Material and Supporting Wooden Flats

1. Check all fill material. It should be cleaned thoroughly; but shall not be displaced from position, and no material shall be lost during cleaning.
2. The loss of material should be made up by addition of fresh fill material.
3. Inspect the support grids for tower fills. Apply fresh rust-resistant coating if some repairs are carried out.

6.5   Conventional Refrigeration Plants

(Refer Appendix D Tables D2 and D3, for commonly used brines and refrigerants respectively).

6.5.1   Some Typical Components Are

- Compressor for the refrigerant.
- Condenser for the refrigerant vapour.
- Cooling tower and water pumps, and their accessories
- Receiver for liquid refrigerant
- Expansion-cum-control valve
- Chiller [evaporator]
- Chilled water tank/brine tank and circulation system
- Cold insulation
- H.P. and L.P. refrigerant piping
- Piping for chilled water/Brine
- Instrumentation and safety valves
- Electrical motors and systems

6.5.2   Observations to Be Made for Safe and Smooth Working

- Compressor: Pressures of refrigerant at suction and discharge of compressor, compressor capacity in use, oil pressure, and oil level in sight glass/in indicator.
- Any vibrations or abnormal noise from compressor.
- Condenser: (1) Pressure of refrigerant vapour at inlet of condenser and (2) pressure and temperature of cooling water at inlet and exit of condenser.
- Flow of cooling water through condenser returned to the cooling tower/ spray pond.

- Any gas leak in the system. Appropriate gas detectors should be installed around the compressor, condenser, liquid receiver, evaporator, etc.
- Also check regularly around various units by portable gas leak detectors. *This is very important for gases like Freon which do not have detectable smell and can be dangerous when inhaled.*
- Insulation on cold surfaces, chilled water tank, and liquid refrigerant piping.
- Chiller [evaporator]: Flow of chilled water through the unit (if possible, to observe visually), pressure of chilled water at inlet, pressure of refrigerant in the unit.
- Pressure of refrigerant at inlet and outlet of chiller.
- Level of refrigerant in liquid receiver and chiller.
- Try to operate at as high a temperature of the circulating chilled water/brine as possible to reduce power consumption by the refrigeration system. This shall be attempted without affecting performance of the main process units.
- Working of all chilled water, brine, and cooling water circulation pumps.
- Analysis of circulating chilled water, brine (pH, total dissolved solids (TDS)).
- *(add anti- corrosive chemicals to the brine if necessary)*
- Analysis of cooling tower water (TDS, pH, suspended matter).
- Some water from cooling tower basin should be drained out regularly in case of TDS reaching the limit (generally not more than 200 ppm in order to minimise scaling on heat transfer surfaces of condenser tubes).
- Level of water in cooling tower and availability of fresh make-up water.
- Safety valve settings on all high pressure units.
- Safety alarm settings/interconnections in case of high refrigerant discharge pressure; very low chilled water flow, low flow of cooling water through condenser, etc.
- Thermal insulation on units and piping operating at low temperature.
- Liquid receiver: Should have well protected level indicator with automatic closing arrangement for connecting (isolation) valves in case of a leak of the indicator.
- Working of control valve for liquid flow to evaporator (for automatic adjustment of refrigerant flow as per cooling requirement and operating level in evaporator: - Very high level of liquid refrigerant in evaporator must be avoided as the liquid can enter the suction side of the compressor.
- **Load on electrical motors** should be monitored for (but not limited to these) compressor motor, cooling water pump, ID fan of cooling tower, and chilled water/brine pump. High and low current alarms shall be installed for all these motors.

6.5.3   Safety Valves

Safety valves should be available for the following units. Their settings should be approved by statutory inspectors.

- Compressor discharge
- Oil separator vessel
- Inlet of condenser
- Inlet of evaporator

6.5.4 Safety Trips/Interconnection

Safety trips/interconnection for tripping of the compressor motor should be provided for the following conditions. The satisfactory working should be checked regularly and recorded. Any fault must be rectified immediately.

- High compressor discharge pressure (some likely reasons: cooling water flow may be less, temperature of cooling water may be high or heat transfer surfaces have become dirty/scaling has taken place in condenser).
- Low lubricating oil pressure (problem with lube oil pump/frothing in oil due to dissolved gas).
- Low cooling water pressure at condenser inlet (due to less supply from cooling water pump).
- Low flow of chilled water through evaporator.
- If the temperature of cooling water at exit of condenser is high (due to low flow).

6.5.5 Some Typical Precautions for Maintenance Work

- Evacuate the refrigerant completely from the unit to be attended and transfer to other unit or to cylinders. Flush out the residual refrigerant by nitrogen and check the exit gas stream by gas detector before taking up maintenance work.
- Arrange good ventilation by a big air circulation fan.
- Proper face shield, safety eye goggles, and full length hand gloves shall be worn.
- Portable breathing apparatus and gas masks should be available.
- Condenser: The tubes are to be cleaned by brush and then washed by clean water. The shell side is to be cleaned by chemical cleaning and washed by clean water. After this both sides are to be tested at a pressure at least 1.5 times the maximum working pressure. Now evacuate all air inside before charging in fresh refrigerant.
- Compressor: To be overhauled as per instructions from Original Equipment Manufacturer (OEM). Any faulty component must be replaced by spare from OEM.
- Check drive motor, cover on the motor, base plate, and alignment with compressor.
- Check and lubricate the bearings.
- Take out oil sample from the crank case (where necessary) and check for any turbidity.

- Evaporator: Stop circulation of chilled water. Evacuate the refrigerant completely and transfer to other unit or to cylinders. Clean the tubes by brush and then wash by fresh water. Check the gaskets (at partition plate in end cover) in a multipass system.
- Receiver for liquid refrigerant: Check the level indicating gauge glass, the connecting nozzles, and isolation valves.
- Check flow control valve for liquid refrigerant.
- Check insulation on liquid receiver, evaporator, and cold pipelines.
- All pressure parts and pipelines are to be tested at 1.5 times the maximum working pressure or as directed by statutory authorities.

6.5.6 Additional Safety Precautions

- Do not expose cylinders (filled with refrigerants) to direct sunlight or temperatures more than 45 °C. Keep them in well-ventilated sheds.
- Liquid ammonia can cause severe cold burns, hence take care not to come in contact.
- Wash with copious amount of water in case of contact.
- Ammonia is also inflammable and can explode if concentration in air is more than 11%.
- Evacuate the system completely by removing all refrigerants before taking up any maintenance work. Bring the pressure to zero and flush out by nitrogen before opening any unit.
- Test the units, pipelines, etc. by compressed nitrogen after the maintenance work is over.
- Appropriate gas detectors should be available to detect leak of refrigerant. *In case of ammonia a wet cotton swab with HCl can indicate the leaking spot of ammonia by generation of fumes around it. A piece of burning sulphur (which produces SO_2 gas) can also indicate source of leak of ammonia by appearance of fumes.*
- Gas masks, face shields, and eye protection safety goggles, portable breathing apparatus, and oxygen cylinders must be available for protection.

6.6 Vapour Absorption Chiller (Fig. 6.1)

This works on the principle that water will boil at a very low temperature when under high vacuum. This system can be installed in a plant which requires chilled water at about 8–10 °C and has low pressure steam or hot water available.

A general description is as follows:

- Water in a section of the equipment is made to boil at a low temperature of about 4.5–5.0 °C by maintaining it under high vacuum. Incoming chilled water (at a higher temperature) from process units is passed through coils kept immersed in this boiling water. The incoming chilled water gives heat to the water boiling at low temperature and therefore gets cooled. It is then sent to the process units.
- The vapours of the boiling water are absorbed in cooled concentrated solution of Lithium Bromide (LiBr) and vacuum is thus maintained. The LiBr solution gets diluted in the absorption section of the equipment.
- The dilute LiBr solution is then sent to an evaporator for concentration. Low pressure (LP) steam or hot water are used as source of heat for evaporation.
- Water vapour produced during the concentration of LiBr solution is sent to a condenser and the condensed water is returned to the boiling section.
- Water thus works as refrigerant in this system.
- The concentrated LiBr is cooled (by dilute LiBr solution which is being sent to the evaporator for concentration) before it is used to absorb the water vapour produced in the boiling section.
- Power is required only for pumping the dilute LiBr solution to the evaporator. There is no compression of vapours of refrigerant.

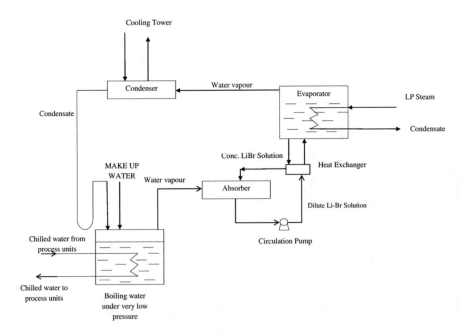

WORKING PRINCIPLE OF VAPOUR ABSORPTION CHILLER

Fig. 6.1 Vapour Absorption Chiller

- This method thus has the advantage of low power consumption and use of LP steam. *However, it cannot produce sub-zero temperatures if brines are to be circulated.*
- Care should be taken to control the steam pressure and to maintain the cooling water flow through the condenser.
- Vacuum in the system needs to be maintained in the boiling section for proper cooling of the chilled water passing through the coils immersed in the boiling section. Hence all joints in the equipment should be leakproof.
- Care shall be taken while handling LiBr due to its corrosive nature.

There are some well-known manufacturers of such machines. They have many models to choose from. Readers may contact Kirloskar, Carrier, Hitachi, Thermax for further information.

6.6.1   Check Points for Proper Working

Following general guidelines are given for working of the vapour absorption chiller. It is suggested that instructions from the manufacturer for safe and smooth working of the units must be followed.

- Working of vacuum pump for creating vacuum at start.
- Operating pressure (vacuum) in the plant. Provide a sensitive gauge for monitoring the very low pressure to be measured.
- Working of steam pressure regulator for controlled heating of LiBr solution in the evaporator.
- Cooling water flow through condenser.
- Water level in boiling section.
- Temperature of the boiling water.
- Make up water quality for the boiling section (very little quantity will be normally required). This shall be demineralised water preferably.
- Check density of LiBr solutions (dilute and concentrated).
- Temperatures of dilute and concentrated LiBr solutions at inlet and exit of the heat exchanger.
- Ensure proper flow of chilled water through the coils immersed in the boiling water.
- Monitor pressure and temperature of chilled water at inlet and outlet.
- Additional check points as per recommendation of manufacturer.
- In a variant of this design, NaOH solutions are used instead of LiBr solution. It is generally a tailor-made design as per clients' requirement.

6.7 Heating Systems

Various methods are used for heating the process units, liquids, and gas streams in chemical industries by electrical means, by steam at appropriate pressure, by firing of fuels like oil and coal, by hot process gases, by heat transfer oils, etc. Some important criteria for selection of the method are given below:

6.7.1 Considerations for Heating Duty

• Heating load i.e., Total Kcal required. (Consider quantity of material to be heated)

Any phase change of the material (in a Melter, evaporator) has a higher heating load as latent heat of melting/evaporation is also to be supplied.

• Rate of heating required Kcal/hour.
• Initial temperature of material/equipment to be heated.
• Final temperature to be achieved.
• Properties of material to be heated (specific heat, melting and boiling points, flash point, viscosity, moisture content).
• Accuracy of temperature control required during heating.
• Whether heating is to be done continuously or intermittently.
• Cost of heat available for use Rs/Kcal (electrical heating is costliest while heating by hot process gases can have very low operating cost).

6.7.2 Physical and Chemical Properties of the Heating Medium/Fuels

(Refer Appendix D-4 for properties of fuels)
    Fuels are used in process plants for open fired heaters, reactors, open fired evaporators, kilns, steam generators, gas turbines, and IC engines. Commonly used fuels are petroleum liquids, natural gas, and solid fuels such as coal, coke, and waste products. Following properties should be known while selecting the fuel.

• Net and gross calorific values
• Viscosity at various temperatures
• Melting point, boiling point, and flash point
• Moisture, sulphur, and ash content
• Specific gravity
• Physical state in which the fuel is available (as big lumps, as small-sized pieces, as a thick viscous liquid, as a gas under moderate pressure).

6.7.3 System Design Criteria

- Design and heat capacity of the process vessel/equipment used for heating.
- Orientations of incoming and outgoing streams to and from the process vessel.
- Whether the material to be heated is very sensitive to higher temperatures. The higher temperature limit shall be accurately known. In this case the temperature of heating medium needs to be limited; and co-current heating may have to be done.
- Whether heating is to be done quickly i.e., whether high rate of rise of temperature is required with accurate control of temperature of material to be heated.
- Whether large amount of heat is to be supplied quickly.
- Whether the material is to be melted or concentrated by evaporation of solvent from the solution. In these cases, the latent heat is also to be supplied in addition to sensible heat.
- If the material/equipment is to be only kept hot, it will be necessary to only make up the losses of heat to the surroundings.
- Range in which the temperature of the materials is to be controlled (whether within ±5–6° OR within ±20°) accurately.
- *Electrical heating and oil/gas firing can control temperature more accurately.*
- Instrumentation required: Appropriate thermocouples for the type of heating system used; pressure and flow controllers for fuel and material to be heated; alarms for high temperatures and safety interconnections/trip systems.

6.7.4 Practical Guidelines for Heating Systems

Electrical Heating

By this method high temperatures can be achieved quickly with accurate temperature control. It needs step-down transformer of adequate capacity, with arrangements for on-load tap changer for smooth step-less control of heating; bus bars capable of carrying high current; refractory lined furnace with provision of graphite electrodes (generally); special insulating supports for bus-bars; ammeters, voltmeter, KA, KV, KW, and KWH meters, and emergency trip system and suitable circuit breakers/switch gear.

It also needs a barricade type enclosure around transformer, main circuit breaker, and metering panel.

There are chances of fires due to short circuits. Hence firefighting arrangements are necessary (especially for arresting oil fires) near transformer, control panel, switch gear, etc.

Plant engineers should note that the cost of power to generate heat is also considerably more as compared to other methods.

Since no ash or smoke is generated, pollution control equipment are not required.

Electricity may be purchased from external grid supply if the cost is less than cost of own generation by diesel engine driven generators.

Steam Heating

Saturated steam can supply large amount of heat; and hence a lesser heat transfer area may be required (*Superheated steam is generally used for power generation*).

However, this method can heat the process vessel up to a temperature corresponding to steam pressure only. This method is generally used for process materials which can get damaged at higher temperatures.

It can be an ideal choice when low pressure steam is available from exhaust of steam turbines. Hence cogeneration shall be considered by plant designers.

This method needs a boiler (with economiser) capable of generating sufficient steam at required pressure, handling system and firing arrangement for fuel, water treatment equipment and feed water pumps, Pressure Reducing Station (Valve) for steam, suitable steam traps, thermal insulation and cladding for piping, and a condensate recovery system.

Steam heating system consists of pressurised components and needs approval from statutory authority. Ash handling arrangements will be required if coal is used as fuel.

Heating by Hot Thermic Fluids (Heat Transfer Oils)

- Can achieve higher temperatures (290 ° C) without pressurised jackets for the process vessel to be heated. Needs thermic fluid oil heating and circulation system with expansion tank.
- Can be quickly started and stopped.
- Needs greater heat transfer surfaces as compared to heating by condensing steam due to lower heat transfer coefficient.
- Should be designed for fire hazard from circulating thermic oil streams. Piping, flanges, and gaskets in pipelines should be selected accordingly.
- The pumps should have mechanical shaft seals.
- The motors and electrical fittings shall be flame proof type.
- All piping, valves, etc. shall be tested at a pressure of 8 kg/cm^2 at least before commissioning.
- The thermic fluid oil is generally heated by fuel oil firing.
- Proper shed to protect from rains should be provided.

Coal Fired Systems

This system needs a coal storage, weighing, conveying, crushing/pulverising and screening arrangements, dust suppression and handling system, screw feeders and burners for pulverised coal; primary and secondary supply air lines with forced draft FD and induced draft ID fans, air pre-heaters, air pollution control facilities, and tall stack (chimney).

It takes more time to start and reach operating temperatures. It also takes more time to cool down if heating is not required. Hence this system is not suitable for quick start and stop requirements.

Arrangements are also required for safe collection and disposal of the ash generated when coal with high ash content is used. This can be a major problem if large quantities of coal are used every day. However, the use of coal is being reduced at locations where oil and natural gas are available through pipelines.

Oil Fired Systems

This system needs storage tanks with adequate protective devices (lightening arresters, fire-fighting systems), oil transfer pumps, day oil tanks, oil feeding/metering pumps, strainers, heaters for oil piping, oil burners and refractory blocks for fitting the burners in combustion chamber, combustion air blowers, and compressor for atomising fuel oil, insulation for oil pipes, etc.

The oil pipelines and valves should be tested at 8 kg/cm^2 before commissioning. The oil feeding pumps must have mechanical shaft seals and an in-built pressure release valve for safety. Flame monitoring system should be provided. It should stop the oil feed pump in case the flame gets extinguished or air supply pressure gets reduced below set point.

High temperatures can be achieved by oil firing. Good temperature control is possible. This system is suitable for quick start and stop requirements.

Other arrangements required for this method:

* Statutory permission for storage of fuel oil.
* Weigh bridge to monitor incoming supply by tankers.
* Storage tanks (should have a capacity sufficient for 15 days' consumption) equipped with level indicators.
* Dyke walls around the storage tanks to collect any overflow of oil during filling up.
* Lightening arresters
* Firefighting arrangements
* Ramp for incoming tankers (to enable emptying completely). *This is optional.*
* Pumps for transferring fuel oil from supply tankers to the storage tanks, and from storage tanks to the day tanks in the plant.
* The motors and light fittings shall be flame proof.

Gas Fired System

It needs storage tanks for LPG/propane with adequate protective enclosures and devices (lightening arresters, firefighting arrangements), gas piping with insulation, flow meters, pressure regulators, venting system through auxiliary burner and chimney, and flame arrester. A photoelectric flame monitor and rupture disc for the furnace is also advisable.

This system is suitable for quick start and stop requirements. Good temperature control is possible. Gas flames are clean and propagate rapidly burning satisfactorily at very low fuel rates.

The gases are stored under pressure and hence statutory permissions are necessary.

Since no ash or smoke is generated, pollution control equipment are not required.

Heating by natural gas: In certain locations natural gas is available through a piped system. Special meters for measuring the consumption are fitted. This is a clean source of heat and on-site storage is not required. A flare tower may also be provided if the gas is to not be consumed in the plant due to some emergency.

Heating by Hot Process Gases
This can be used in certain chemical plants where hot process gases are available and are to be cooled before further processing in downstream units. Plant engineers shall carefully consider the heating load, temperatures to be achieved after heating, corrosive (depending on the presence of acidic gases and the dew point) and erosive properties of the process gases, and pressure drop allowed for the gases during operations. It should be noted that the heating may get interrupted during any plant stoppage or if higher temperature process gas is required in downstream units.

This option is suitable when temperature of these gases is high enough and heating by other means is not feasible technically or economically.

Initial investment will be needed for equipment for heating by gas; gas ducts for inlet, outlet, and bypass with butterfly valves (manual or motorised), structural supports, instrumentation, thermal insulation of all the gas ducts, etc.

6.7.5 Air Pollution Control Equipment

- There will be need for air pollution control equipment; ash collection and disposal arrangements depending on the sulphur and ash contents when using coal firing system. The flue gases may be corrosive or erosive or both. The MOC of downstream units will have to be chosen accordingly. High ash content in flue gases will need electrostatic precipitators, bag filters, etc.
- Flue gas scrubbing system may be required if a fuel oil with high sulphur content is used.
- Air pollution control equipment are generally not required for heating by LPG, propane, or hot process gases.

6.7.6 Storage and Handling Arrangements

- For solid fuels: Storage yards, weighing and conveying, crushing, pulverizing and screening equipment, silos, and screw feeding. Solid fuels can also be transported within plants by conveyors, buckets and elevators, lifts, and pneumatic systems. The quantity to be transported and the distances involved shall be considered for selecting the mode of transport.
- For liquid fuels: Storage tanks, day tanks and feeding pumps, piping system with heaters for the liquid, strainers, feed control valves, lightening arresters and oil burners, combustion air blowers, and air pre-heaters. The heavy industrial fuels

are sometimes available in steam-heated tankers. The fuel is stored in heated tanks and in large installations circulated continuously between heated storage and furnace. Thus, the fuel is kept at the correct temperature for transfer to the day tanks for burners so that control of firing will be simple.

- For liquefied gases/gaseous fuels: Storage tanks for LPG and propane, pressure regulators, flame arresters and flow meters, burners, lightening arresters and gas burners, combustion air blowers, and air pre-heaters.
- Firefighting arrangements and suitable cordoning of the storage areas must be provided for all areas where fuels are handled.

6.7.7 Manpower Required

Qualified, trained, and licensed manpower is required for operating and maintaining the heating systems. These jobs shall not be given to untrained or semi-skilled persons. Typically, there will be need for trained persons to handle coal pulverisers and feeders, to receive fuel oil from supply tankers, licensed electricians, licensed boiler attendants, process operators, and effluent treatment plant operators.

Trained maintenance personnel will also have to be employed.

References/for Further Reading

1. "Process Equipment Procurement in the Chemical and Related Industries", Kiran Golwalkar, Published by Springer International Publishing, Switzerland.
2. "Production Management of Chemical Industries", Kiran Golwalkar, Published by Springer International Publishing, Switzerland.

Chapter 7
Cogeneration of Steam and Power, Steam Traps, and Heat Exchangers

Abstract Many chemical plants need steam for heating. It is generated either by operating independent boilers or making available through waste heat recovery boilers which recover heat from hot process gases. The steam can be used both for generation of power and process heating. This increases the overall efficiency because latent heat of the exhaust steam from the steam turbines is used for process heating instead of condensing it by cooling water.

Process heating is done efficiently by allowing only the drainage of condensate from the unit being heated while preventing escape of live steam. The condensate is to be recycled to boiler feed water tank. Mechanical, thermostatic, and thermodynamic type of steam traps are available from which a proper choice shall be made for the steam-heated equipment. Selection of steam traps mainly depends on type of heating requirements and generation of condensate. It is necessary to monitor the working of traps.

Plant operations involve heating and cooling of process streams for which appropriately designed and fabricated heat exchangers are necessary. The efficient operation of chemical plants depends on these heat exchangers to a large extent.

Keywords Precautions during steam generation · Waste heat recovery boilers · Simultaneous generation of power and steam · Steam super-heater · Steam trap types · Inverted bucket · Float type · Thermostatic · Thermodynamic traps · Trap selection for operation with superheated steam · Condensate loads · Heat exchangers

7.1   Introduction

Steam is a very useful heating medium in a process plant. It can be generated by burning fuels or by recovery of heat from hot process gases available in the process plant when temperature of the gases is to be brought down *to suit the requirement of downstream units.*

In a large chemical premise, a central boiler house can supply steam for entire plant or smaller boilers may be located near the area of use. Careful technical and economic study should be made to determine the type, capacity, and number of boilers required and most advantageous arrangement of the distribution system.

Chemical engineers should look into the possibility of steam generation through waste heat recovery and simultaneous generation of power (by using suitable steam turbines) to reduce dependence on external source of power wherever feasible.

It is very necessary to take care of important operational matters for safely and efficiently operating the steam generating units as given below.

7.2   Generation of Steam

- Compliance with statutory instructions at all times is a must. There should be no tampering to settings of safety valves by anyone after they are fixed by boiler inspection authority. Only qualified and licensed boiler attendants shall be employed.
- Ensure sufficient stock of treated water always in the boiler feed water tank in the plant. It must be enough for at least 8 h of operation. The water treatment units like softener, RO plant, and demineralisation system etc should be kept in working condition always. Resin columns must be regenerated well in time to ensure availability of treated water all the time. *It should never be required to feed untreated water to the boilers due to non-availability of treated water.*
- All available condensate from process units, evaporators, steam turbine exit, etc. must be recovered and recycled to boiler feed water (BFW) tank to minimise the load on water treatment units.
- Quality of BFW must meet the specifications for the boiler as specified by the boiler manufacturer. This should be continuously monitored by on-line instrumentation.
- Dissolved oxygen, pH, and silica content in the feed water must be very carefully checked. These should always meet specifications given by boiler manufacturer.
- De-aerator operation shall be closely monitored and its temperature may be increased (if necessary) to remove dissolved oxygen.
- Working of feed water stop valves, non-return valves and blow down valves on boilers and economisers must be checked regularly.
- Working of safety valves shall be checked without disturbing the setting.
- Samples of feed water and boiler blow down water must be checked regularly.

- All BFW pumps must be in working order always. Running and standby pumps shall be operated alternately as far as possible.
- Water level controller of the boiler must be in proper working order. Gauge glasses of the boiler must be kept clean always.
- Proper working of steam traps on process heaters and steam jacketed lines must be ensured by monitoring the incoming steam line pressures and downstream condensate flow in the return piping to BFW tank.

7.3   Precautions During Steam Generation

- The quality of fuel must meet specifications since higher ash and sulphur content can cause loss of efficiency and corrosion as well as environmental pollution. This will need electrostatic precipitator and scrubbing arrangements to control the pollution.
- Coal crushing, screening, and screw feeding arrangements should be closely monitored. Oversized coal particles should be recycled for pulverising.
- Any loss of refractory bed particles should be immediately made up by adding high alumina screened particles of appropriate size (generally 5–8 mm) for maintaining proper combustion in the circulating bed.
- Flame monitoring device and alarm (if the flame gets extinguished) must be in working order always.
- Standby oil firing burner must always be in a ready-to-start condition.
- In case of oil-fired boiler the atomisation of fuel oil must be ensured. Flue gases should be checked for presence of soot particles which indicates incomplete combustion.
- Oxygen and CO_2 content in flue gases must be monitored on-line. They should meet specifications given by boiler manufacturer.
- Ensure that the fuel oil is at lower viscosity by checking the working of oil heater.
- Clean/change the burner nozzle if the oil spray is not proper. Check the strainer in oil feeding line for any damage to the screen.
- All fans (FD, ID, and combustion air) and dampers in air lines must be in working order. External indicators on the dampers must indicate the correct position of flap; and it should be possible to lock the operating handle/lever in any desired position.
- Alarm for low air pressure should be in working order.
- Thermal insulation and cladding on all hot surfaces should be checked for any leaks through which ingress of rain water can occur.
- Thermal imaging camera and hand held optical pyrometer shall be used to monitor heat losses.
- Audio-visual alarms and tripping interconnections for low water level in boiler drum, outage of flame in combustion chamber, etc. should be regularly checked.

- During maintenance of the boiler, the moisture separator/demister pad provided for minimising carryover of water droplets (with the outgoing steam from the boiler) should be thoroughly inspected and attended to.

Precautions for Waste Heat Recovery Boilers (WHRB)   Provide electrical interconnections to the process blower to stop the flow of hot process gases through the WHRB in case of low water level. This can be made possible by tripping of process gas blower through the low water level interconnection.

- Operation of butterfly valves in the gas ducts should be checked to suit process conditions.
- Acidic corrosive condensate from the gas side must be drained out regularly.

7.4   Simultaneous Generation of Power and Steam for Heating

Power is generated by passing high pressure (HP) steam through a steam turbine which rotates the rotor of the turbine. This in turn rotates the armature of a generator to produce electricity.

The difference in energy content of the steam at inlet and exit of the steam turbine is available for generation of power.

Steam from the exit of the turbine is condensed by circulation of cooling water through the condenser to maintain very low pressure (vacuum) in the turbine. The condensate is returned to the boiler which is generating the HP steam.

Since the latent heat of condensation is quite high a large proportion of the input energy is *lost to cooling water* when the exit steam is fully condensed when this type of turbine is used. Hence in many chemical industries the exit steam from the turbine is used for process heating whereby the latent heat is used (recovered) (Figs. 7.1 and 7.2).

This increases the overall efficiency of the system because the energy available in the input steam is used for both power generation as well as process heating. It also reduces the cooling load on the cooling towers which supply cooling water for condensation of steam.

The pressure at which steam is to be taken out from the exit of the turbine (either back pressure type or extraction type) depends on the temperature at which process heating is to be done. The heat required for the process is also to be taken into account.

Example   In a chemical process plant steam is generated at a pressure of 40 kg/cm^2 by the WHRB boiler and superheated to 400 °C. It is then admitted into the turbine. The exhaust steam is taken out at 6.0 kg/cm^2 and the latent heat is used for process heating. This increases the efficiency of the system by generating power as well as using the exhaust steam for process heating.

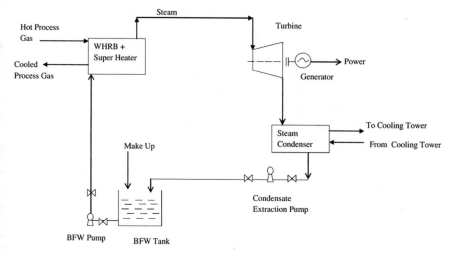

LATENT HEAT OF STEAM IS LOST TO COOLING WATER

Fig. 7.1 Heat lost when exit steam from turbine is fully condensed

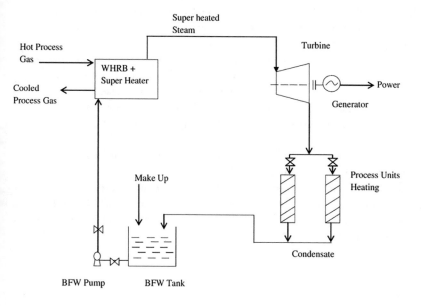

LATENT HEAT OF STEAM IS USED FOR PROCESS HEATING

Fig. 7.2 Exit steam from the turbine used for process heating

However, if it were taken out at 0.1 kg/cm^2 from the turbine exit and condensed fully by a water-cooled steam condenser, the latent heat would have been lost to the cooling water.

Plant engineers shall look in to the properties of steam and condensate at different pressures and temperatures (such as enthalpy and latent heats) and estimate the enthalpy difference available for power generation as well as the latent heat available for process heating per MT of steam passing through the back pressure turbine.

Various combinations of the pressure and degree of superheat at inlet of turbine and the saturation temperature corresponding to exhaust pressure can be considered to meet the requirements of power and heat.

Calculations shall be made to use superheated steam at inlet of turbine for power generation and exhaust from the turbine exit at a pressure (and corresponding temperature) for process heating so that maximum energy is utilised.

7.4.1 Steam Super-Heater

Moisture in steam is undesirable for most systems as wet steam can erode turbine blades.

The heating value can also be less to the extent of moisture present in the steam.

Therefore, the saturated steam is further heated above its saturation temperature by passing through a heat exchanger called super-heater which is located inside of the furnace or a suitable source of heat.

This ensures that only dry steam is supplied to the turbine. Hence power generation can be increased if it is possible to superheat the incoming steam through a suitable source of heat in the plant. However, excessively superheated steam can be harmful to the steam turbine blades. Hence a de-superheater shall be provided for controlling the temperature of the steam by spray of demineralised water at controlled rate.

The manufacturer of steam turbine shall be consulted for the maximum permissible temperature of steam at inlet of turbine.

The boiler and super-heater shall be specified and procured accordingly to generate high pressure superheated steam. The backpressure turbine shall be selected to suit power and heating requirements in the plant accordingly. Future needs of power and heat shall also be considered.

7.5 Steam Traps

Introduction

Steam is an important heating medium for heating of process reactors, for evaporators, for melters, and for keeping process units and pipelines hot. It is necessary to prevent loss of live steam for maintaining the efficiency of heating. Steam traps are therefore provided for allowing only the condensate to pass through without allowing live steam to escape. The steam traps must also allow discharge of non-condensing gases such as air which could be present initially in the system.

The operation of a steam trap depends on the difference in properties between steam and condensate. The liquid condensate has a much higher density than steam and hence it tends to accumulate at the lowest possible point in the system.

Steam traps have three main categories, mechanical, thermostatic, and thermodynamic, which work on different operating principles to remove condensate and non-condensable gases and prevent loss of steam from the system.

Mechanical Traps These remove condensate by using the difference in densities of steam and condensate. The liquid condensate goes to the lowest part of the system. Mechanical traps have a bucket or a float that rises and falls in relation to condensate level. They usually have a mechanical linkage attached that opens and closes the exit valve as per condensate levels present in the body of the steam trap. Inverted bucket and float traps are the main types of mechanical traps.

7.5.1   Inverted Bucket Type

These traps have an inverted bucket within the trap which is attached to a lever that opens and closes the exit valve in response to the motion of the bucket. When steam or air flows into the underside of the inverted bucket and condensate surrounds it on the outside, the steam causes the bucket to rise. In this position, the bucket will cause the exit valve to close. There is a vent hole in the top of the bucket that allows a small amount of the vapor to be released into the top of the trap, from where it is discharged downstream. As vapor escapes through the vent hole, condensate starts to fill the inside of the bucket, causing it to sink. This causes the attached lever to open the trap valve and discharge the condensate (along with any vapor held in the trap) (Fig. 7.3).

7.5.2   Float Type

Float traps have a sealed spherical float inside. They have a mechanical linkage to operate the exit valve or can operate the exit valve through use of float itself. The position of the float changes according to the level of condensate in the trap. Movement of the float opens/closes the exit valve.

There are two basic designs used for float traps: lever float and free float.

Lever float type traps: The float is attached to a lever which can operate the trap exit valve. As condensate enters the trap, the float starts rising and moves the lever, causing the trap valve to open.

Free float type traps: In this design, the float is not attached to a lever, and the float itself serves as the valve for the trap. A free float is able to independently rise

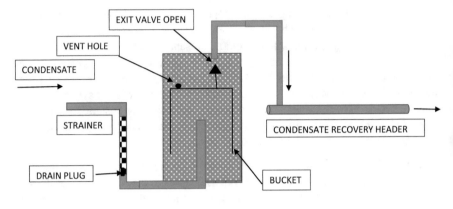

BUCKET AT LOWER POSITION

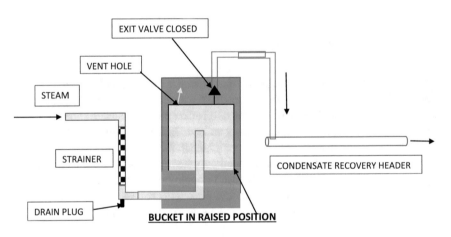

BUCKET IN RAISED POSITION

Fig. 7.3 Working principle of inverted bucket steam trap

away from the exit orifice allowing the condensate to be drained out. There is no obstruction to the flow of condensate).

Continuous Drainage: A Significant Advantage of Float Type Traps

One key difference in the operation of float traps and inverted bucket traps is the type of condensate drainage they provide. While float traps provide continuous drainage, the inverted bucket traps provide intermittent drainage.

The float rises and falls depending on the level of condensate in steam traps that continuously drain condensate. When condensate enters, the valve opens just enough to drain the condensate, closing the exit valve after the flow of condensate ceases. This allows the trap to respond quickly to fluctuations in condensate load.

On the other hand, inverted bucket type steam traps drain intermittently. The condensate is not drained until a significant amount of vapor is vented from the bucket, thereby causing the bucket to sink and the valve to open.

The generation of condensate from steam-heated equipment is generally contin-uous and needs to be removed as soon as it is formed. Hence one should provide a trap which drains out the condensate continuously. However, the condensate will accumulate within the trap or the system as long as the exit valve remains closed in case of steam traps which drain intermittently.

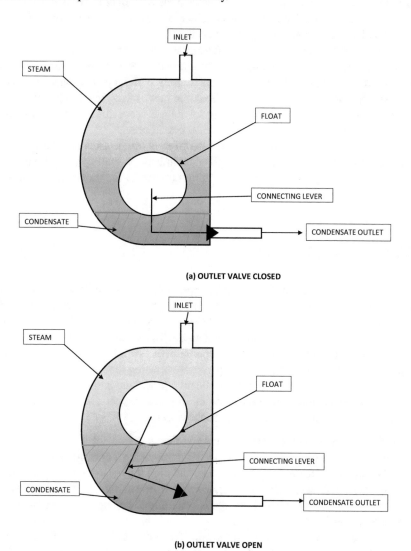

Fig. 7.4 Working principle of float type steam trap (**a**) Outlet Valve closed (**b**) Outlet valve open

7.5.3   Trap Selection for Operation with Superheated Steam

In systems operating with superheated steam, there is often little condensate. During such operations, there may not be sufficient water inside an inverted bucket trap to float the bucket. As a result, the bucket sinks to the bottom of the trap and the outlet valve remains open causing loss of considerable amounts of superheated steam. This can also exert higher backpressure on the trap.

Lever float traps are also affected during use in superheated systems. In **lever float** traps, the valve head is very close to the seat. Low flow operation may cause condensate to flow through the valve at extremely high velocities, causing erosion of valve components. However, a **free float** can move away from the top of the seat during low flow of condensate. Since the valve head is not directly in the flow path, the erosion is prevented even under low flow conditions.

Mechanical traps are manufactured with different orifice sizes based on pressure differential available for each model. The manufacturer shall be consulted during selection of the trap for the given operating conditions for proper functioning of the trap.

7.5.4   Thermostatic Traps

These remove condensate through the temperature difference between live steam and the condensate. The exit valve operates through expansion and contraction of a mechanism (which consists of a filled element/bellows or a bimetallic element) that is exposed to the heat from steam or condensate. These traps open at the lower temperature of the condensate below the temperature of steam and allow the condensate to pass through. Thermostatic traps operate by utilising the difference in temperature between condensate that is close to steam temperature and sub-cooled condensate (or low temperature air). The operating mechanism needs regular maintenance of the bellows/bimetallic element for proper working.

Typical Applications of Thermostatic Type Traps

Free float steam traps are generally used for removing condensate from steam-heated equipment, steam distribution lines, and high temperature tracing *where the condensate must be removed as soon as formed*. However, this is not very necessary for tracer lines since the material inside is generally to be kept at a lower temperature than the temperature of steam.

In such applications bi-metal, expansion type, or specially filled balanced pressure thermostatic traps are useful because these traps can reduce the discharge temperature of condensate *below the saturation temperatures*. They can help to use latent heat plus some sensible heat from condensate.

7.5.5 Thermodynamic Traps

There are two main varieties of thermodynamic steam trap: the thermodynamic disc and thermodynamic piston (impulse) types.

However, disc traps are most commonly used since impulse traps can leak pilot steam, and may fail in the presence of even a slight amount of dirt blocking the pilot channel.

Operating Mechanism of Thermodynamic Disc Traps
Thermodynamic disc steam traps have an intermittent, cyclic operating characteristic. These traps have an internal disc, an orifice to release condensate, and seat rings. It opens to discharge condensate for a few seconds; and then closes for a generally longer period until a new discharge cycle begins. The opening and closing action of thermodynamic disc traps is caused by the difference in the forces acting on the bottom and top sides of the valve disc.

Pressure below the disc decreases when steam flows under the valve disc while the pressure within the chamber above the disc is more. The valve disc gets forced onto the valve seat as a result and it closes the valve.

Control steam within the pressure chamber exerts a downward force on the top of the valve disc equal to the product of area and pressure ($A \times P$). Control steam on the underside of the disc causes a pressure drop under the disc because of its high velocity (as long as the disc is in the open position).

Applications where disc-type traps are suitable:
Disc type steam traps are compact and suitable for a wide pressure range. They are commonly used for steam mains drainage or high temperature tracing discharge. *Since free float traps are more energy efficient and durable for the same services they are preferred as a first choice compared to disc type traps in those applications.*

However, there are several applications where disc traps are recommended such as high pressure steam mains drainage service. For this reason, disc type traps are used in severe high pressure conditions, such as with supercritical steam/condensate service.

Additionally, facilities sometimes use disc type traps where there is a risk of freezing. Those disc traps are installed in a vertical down position, as condensate does not pool inside the body when installed in this position, thus the trap is unlikely to freeze. However, this type of installation can need frequent maintenance due to frequent cyclic discharge of condensate.

Advantages of thermodynamic disc traps:

- Can operate over a wide pressure range
- Relatively simple construction
- Compact size
- Low initial cost
- Easy to install in vertical or horizontal position due to compact size
- Can be used on superheated steam

- Highly resistant to damage from freezing as very little water remains inside trap body
- Can be used for wide variety of tracing and certain light process steam applications.

Limitations of thermodynamic disc traps:

- They have a shorter service life due to more wear of the disc and seat during operation.
- There can be greater steam loss during operation as there is no water seal

In case of an impulse type trap, the movable piston has a narrow passage for discharge of condensate; while there is a flange on the upper side. As the condensate is discharged the flash steam above the flange causes a force to push down the piston on the valve seat which prevents escape of live steam. However, small dirt particles or scales in the incoming line can clog the narrow passage and prevent discharge of the condensate. Hence it needs an efficient stainless steel strainer in the inlet line (which should be cleaned and maintained properly).

At start-up, incoming fluids consisting of air, and/or condensate (and even sometimes steam) at line pressure exert a lifting force on the bottom of the valve disc, thereby causing it to rise and open to allow condensate flow.

The valve is designed to close on flashing of the condensate near-to-steam temperature, which occurs once the accumulated condensate is discharged. When this closing force is sufficiently more than the opening (lifting) force, the valve closes.

Advanced thermodynamic disc traps:

Certain advanced thermodynamic disc traps have a built-in thermostatic air vent that works to remove air on start-up. Disc traps with this innovative feature are thermodynamic disc traps with thermostatic vents.

On start-up, the thermostatic ring holds the disc off the seat until the air is vented out of the system. As air is released, the fluid temperature rises allowing the thermostatic ring to expand and move downward to the lower resting position.

In certain designs a bellow keeps the air vent valve open till steam/hot condensate comes in. The bellow gets heated and closes the air vent valve. Thereafter the trap works in a normal manner.

7.5.6 Steam Trap Selection for Heating Duties

- Calculate condensate quantities to be removed. Determine discharge requirements of the steam trap application (e.g. whether the condensate is to be removed immediately or at a somewhat lower temperature than condensation temperature to use sensible heat from the condensate also).
- Calculate back pressure on the trap due to condensate recovery system.
- Considerations for trap selection are steam pressure and temperature ratings, discharge capacity, trap type.

- Select trap model according to operating pressure, temperature, orientation, and any other relevant conditions.
- Confirm from the trap manufacturer.
- It is advisable to provide a stainless steel strainer with a strong screen in the inlet line of a steam trap for arresting pipe scales/deposits, etc. This is to prevent clogging of the orifice for condensate exit from the trap.

7.5.6.1 For Steam Distribution Piping

The steam distribution piping is to be used to reliably supply steam to the steam-using equipment. One of the most important roles of steam traps on steam piping is to reduce the occurrence of water hammer. This is done by selecting a trap which does not accumulate condensate in the system. This can be possible with steam traps which should remove condensate as soon as it is formed instead of sub-cooling of condensate.

Typical requirements are:

- Tight seal to minimise steam leakage even with low condensate loads.
- Unaffected by environment, even in adverse weather conditions.
- Ability to vent start-up air and operating air.
- Continuous condensate discharge.
- Should be unaffected by back pressure.

7.5.6.2 For Steam-Heated Equipment

The condensate generated by these equipment needs to be continuously discharged. Such applications also require an air venting function in the trap to remove air and other non-condensable gases trapped in equipment and connected piping.

Typical requirements are:

- Continuously discharge condensate to maintain heating efficiency.
- Minimize hold up of condensate in the system (e.g. heating jacket of melter, evaporator coils).
- Shall remain unaffected by large variations in condensate load.
- Ability to vent both start-up and operating air.
- Ability to discharge condensate even at lowest available differential pressure and operate effectively against high backpressure.
- Even in case of internal damage/wearing out the condensate shall be discharged.

Requirements of traps for tracer lines are:

- The trap should be able to efficiently use the latent heat of steam and sensible heat of condensate by sub-cooling below saturation temperature.
- It should be suitable for operation in all piping orientations.
- Should have compact size.

- Sub-cooling of condensate shall be suitable for maintaining tracer lines at lower temperature required. Consult manufacturer of steam trap for suitable model.

7.5.6.3   For Steam-Driven Equipment

These types of equipment include turbines used for driving compressor, pump, or generator, and may also include steam hammers or wheels. Condensate should be removed as quickly as possible for safe and effective operation of each power-drive application. It should not pool inside the equipment to prevent damage.

Desirable characteristics of steam traps are:

- Tight seal to minimise steam leakage even with very low condensate loads.
- Unaffected by environment, even in adverse weather conditions.
- Ability to vent start-up air.
- Continuous condensate discharge to minimise holding up in the system.
- Should be unaffected by back pressure.

7.5.7   Testing of Steam Traps

Steam is used as a heating medium in many process units like evaporators, melters, and dissolvers, which have higher continuous steam consumption. Certain reactors can also have higher steam consumption to carry out the reactions continuously. These units have heating jackets or coils through which the steam flows. Appropriate steam traps shall be provided for this purpose. It is necessary that the condensate is removed continuously from the units in order to maintain their performance.

In old practice it was possible to easily check the functioning of the steam traps when the condensate was discharged in the open. However, in modern practice the condensate is recovered by connecting the exit piping from traps to a common header (and then sent to a boiler feed water tank). It is difficult to know in such a case which trap is discharging the condensate properly, which one is passing out live steam and which one is not removing the condensate at all.

The "hissing" sound in the exit line from the steam trap is therefore monitored by a special sensitive probe which is held to closely touch the piping. A continuous "hissing "sound indicates that live steam is flowing out continuously.

Agencies having trained specialists shall be engaged for an audit of the traps in a large process plant.

If there is no sound at all it means there is probably a hold up of condensate in the system and the discharge orifice of the trap may be clogged. This can also be confirmed by checking the temperatures of steam line and condensate line. However, it must be considered that in certain cases it is desired to recover part of sensible heat from the condensate also and the system is designed accordingly (*by selecting the traps in consultation with the manufacturer*).

In such cases there could be a *longer interval* of time before the hissing sound is heard again.

When a periodic (cyclic) sound is detected followed by periods of comparative silence (or very less sound) it indicates intermittent discharge of condensate (when inverted bucket type of steam traps are used).

In case when disc type traps are used there will be an almost continuous chattering sound in the condensate exit line indicating continuous removal of condensate (e.g. traps provided on steam coils for melters).

A Crude Practical Test for Traps on Jacketed Liquid Sulphur Piping/ Melter Coil

The steam inlet and condensate exit lines of a trap are checked by rubbing a piece of solid sulphur on them. In case of a trap functioning properly the piece of sulphur will melt on the inlet line but will not melt on the exit line from the trap.

7.5.8   Generation of Condensate Loads

Melters	High load
Evaporators	High loads
For keeping material hot in a vessel	Less load
Steam-driven equipment	Moderate load
Tracing lines	Less load

7.5.9   Some Typical Reasons for Steam Traps Not Working

- High back pressure due to too long condensate recovery lines.
- High back pressure if the condensate is to be put back in tanks at elevated location.
- Dirt, pipe scales in incoming line.
- More than one trap discharging in a common header which is of too small diameter.
- Damaged internal linkage of float/bucket with exit valve.
- Clogged out internal orifice of the bucket/piston.
- Trap is frozen due to extreme cold environment (needs external insulation).

*Major part of the above information is taken from website of **TLV International**. Readers may contact TLV International (who are the manufacturers of steam traps and related items) for more information.*

7.6   Heat Exchangers

Heat exchangers form a very important part of a process plant. Their safe and efficient performance is a must for functioning of the plant. Heat exchangers are used for maintaining proper temperatures of process streams, for cooling of liquids, for preheating combustion air by heat exchange with hot exit gases, etc.

7.6.1   Types of Heat Exchangers

Shell and tube type, disc and donut type; with external heating and cooling jackets; with internal and external heating and cooling coils; with finned tubes; and plate heat exchangers.

The chemical and mechanical engineers should look into the following guidelines:

7.6.2   Design Considerations

- Define required performance as the heat transfer required (heat to be gained by fluid being heated, and heat to be given by heating medium).
- Select heat exchangers by considering the properties of hot and cold fluids such as specific heat, viscosity, thermal conductivity, density, boiling and melting points, pH, concentration of solution being heated/cooled, and presence of corrosive and erosive particles in the fluids being handled.
- Consider solubility of (dissolved) salts at all operating temperatures. This will indicate chances of precipitation on heat transfer surfaces or clogging of tubes.
- Disc and donut heat exchangers: These are shell and tube type units. However, they do not have segmental shaped baffles on the shell side. Instead, the baffles are alternately shaped in the form of discs and donuts. Gases flowing on the shell side take a zig-zag path when flowing across the discs and donuts. These generally do not have idle pockets on the shell side,have lower pressure drop, and better heat transfer. They permit flexibility for layout of gas ducts with respect to nozzle orientation. They are useful for gas-to-gas heat exchange duties.
- Plate heat exchangers (PHE) have high overall heat transfer coefficients and are compact. They need much less space as compared to shell and tube type heat exchangers. However, they have very narrow passages for the flow of fluids and can get clogged. Hence strainers should be provided at entry points of the process fluids.
- Certain pharmaceutical materials are heat sensitive. They shall not be allowed to get overheated. Co-current flow of process fluid shall be found useful for such operations. Arrange the incoming and outgoing flow of materials to be heated and the heating medium accordingly.

- Gaseous process streams having dust particles, acidic droplets shall be put on tube side of a shell-and-tube heat exchanger. It is convenient to clean the tubes by brushing.
- If such process streams are put on the shell side the cleaning is difficult and heat transfer efficiency is reduced. They generally need circulation of cleaning chemicals on the shell side for restoring the efficiency.
- Carefully consider the maximum and minimum temperatures of the process streams (which can occur during operation) and chances of boiling of liquids in the tubes or shell. The vapours can pressurise the heat exchanger if flow of cooling medium reduces due to some reason. Safety vents/safety valves shall be provided for such heat exchangers.
- Both maximum and minimum flow rates of process streams shall also be considered which can occur when the plant is sometimes operated at a little above or below the rated capacity. It should be possible to plug a few tubes when the plant is operating with low flow rates (for getting sufficient fluid velocities in remaining tubes). Such designs can be helpful for plant operations.
- Select materials of construction of tubes, shell, baffles, and tie rods according to the corrosive properties of the fluids being handled.
- Protective lining required for tubes or tube sheets is also to be considered if there are chances of condensation of acidic (corrosive) materials in case moisture is present in the gaseous process streams being cooled. Cast Iron sleeves/gills are generally provided on steel tubes for protection.
- Refractory/acid-resistant ferrules shall be provided at entrance of the tubes. A layer of castable refractory/acid-resistant cement shall be laid in the space between the ferrules to protect the tube sheet.
- Provide expansion bellows to minimise stresses which can occur during operation when large temperature changes are likely.
- Reinforcing pad plates shall be provided at fluid entry and exit nozzles.
- Pressure drop in the unit must be minimised by providing sufficient cross sectional area but *without* too much reduction in fluid velocity (which can reduce the heat transfer coefficient if it becomes too low). Check Reynold's number for fluid flow.
- Check possibility to have more number of passes to accommodate the unit in given space. However, the temperature difference driving force shall also be checked.
- More number of passes can increase the fluid velocity on the tube side and hence the Reynold's number for heat transfer coefficient. But this can increase the pressure drop also.
- Finned tubes may also be considered if there is very little or no chance of deposit of dust particles (in the space between the fins) from the fluid being handled.

7.6.3   Safety and Convenience

- The heat exchanger is to be designed and fabricated by considering it as a pressure vessel when the operating pressure/temperature is high; or when any of the process streams contain dangerous materials. This consideration is from safety point of view.
- There should be provision of draining out any condensate/liquid which may accumulate inside on tube side or shell side passes. Drain valves should be fitted at appropriate places.
- Vent nozzles and valves shall also be provided for controlled release of gases from the unit before taking up maintenance work (or for release of non-condensable gases/air during restart).
- Cleaning manholes should be available on both shell and tube sides. These could be provided with bolted or welded cover plates as per operating conditions and properties of the process streams. Work platforms and railings shall be available wherever necessary for opening the cleaning manholes.
- Orientation of these manholes should be suitable as per space available in the plant layout.
- Nozzles shall be provided for measurement of temperature and pressure.
- Lifting lugs shall be available.
- Cleats shall be welded on surfaces which need to be provided with thermal insulation.

Readers may contact **V K Engineers, MIDC Tarapur, Maharashtra State, India,** for further information on disc and donut heat exchangers

References/for Further Reading

1. "Process Equipment Procurement in the Chemical and Related Industries", Kiran Golwalkar, Published by Springer International Publishing, Switzerland.
2. "Production Management of Chemical Industries", Kiran Golwalkar, Published by Springer International Publishing, Switzerland.
3. "Integrated Maintenance and Energy Management in the Chemical Industries" Kiran Golwalkar; Published by Springer International.

Chapter 8
Process Control and Instrumentation

Abstract Proper control of the operations and various reactions is essential for safe, efficient, and pollution-free working of a chemical plant. This can result in proper conversion of the raw materials to finished products and minimise their wastage. It also minimises breakdowns of process units and plant machinery and improves the product quality and reputation of the organisation.

The process parameters shall be monitored by installing sensors of the instruments at the right place in the process units and piping. Sampling arrangement should be properly designed. Certain practical guidelines are also given for precautions to be taken while providing instruments and connecting cables.

Choice of feed forward or feedback control depends on the knowledge of relationship of control action with the parameter being controlled; upper and lower set points (allowable deviation from required value) and response time of system.

Keywords Objectives of instrumentation and control · Advantages of process control · Typical Hardware and problems with hardware · Sampling systems · Preventive measures for process control · Feed forward and feedback control · Typical examples

8.1 Introduction

It is very important to exercise proper control on the various processes being carried out in a chemical plant to ensure safe, pollution-free, and cost-efficient running.

Instruments are used in chemical industries to measure process variables such as temperature, pressure, density, viscosity, level, and chemical composition etc.

8.2   Objectives That Can Be Achieved by Instrumentation and Control Are

- Safe running of the plant.
- Reduction in manpower by use of automatic process controllers.
- Easy to monitor the process parameter in a central control room in case of a complex process.
- Easy to control plant operations by use of sophisticated control instruments and devices.
- Better control of the process by use of sophisticated control instruments and devices.
- Production schedules can be better since better product quality reduces rejection by customers; hence easier to meet requirements of customers.
- Less wastage of material by better control of process parameters.

8.3   Advantages of Process Control

- Safety for all persons in working areas and outside the premises.
- Safety for plant assets and surroundings.
- Consistent better product quality while also reducing the cost of production.
- Longer equipment life is possible as the process parameters remain within prescribed limits. Wear and tear of process units and machinery is reduced which in turn can minimise expenses for maintenance and inventory of spares.
- Less consumption of raw materials and utilities can result in reduced generation of effluents.
- Better employee morale as the production unit becomes more safe to operate and is less polluting due to better control of process conditions (better control of high temperatures and pressures).
- Proper control of corrosive, inflammable, and toxic chemicals being handled.

8.4   Some Typical Hardware Required

- Sensors and probes for instruments.
- Accurate field and control room instruments for indicating, recording and controlling process conditions.
- Connecting cables.
- Air-drying plant with accessories.
- Pneumatic control arrangements (flapper nozzle, air pressure control).
- Thermocouples, RTD, pressure gauges.
- Electromagnetic systems for control valves.

- Compensating and power cables.
- Cable trays, junction boxes.
- Metering pumps, rotameters and measuring feed system like screw conveyor for input to process units (reactors, evaporators, mixers, filter press, etc.). This is to ensure that feeding of materials is at controlled rate only for proper functioning of the process units. The feed systems should have locking arrangements for the set points.
- Speed control devices (VFD, mechanical fluid couplings, gear-and-belts with pulley, friction drive).
- Pressure control valves (example: for steam flow to heating jackets of reactors).
- Flow control valves for supply of material on input side (typically operated by level, temperature, pH of downstream process units).
- Limit control switches/mechanism for levels, weights, pressure, flow rates, pH etc

8.5 Design of Sampling Systems

The samples must represent actual conditions of the process streams:

- Sampling lines must be short and should be always kept clean and flushed by process stream/by dry nitrogen.
- Special filters and cooling arrangements should be provided in the sampling lines for proper conditioning of samples.
- Waste from these lines shall be collected, neutralized, and disposed of safely.
- Make provision for collection of multiple samples from reaction units, incoming and outgoing process streams through more than one location for confirmation of analysis.

8.6 Typical Problems with Hardware

- Air for operation of controlling instruments is not of proper quality or not at sufficient pressure due to malfunctioning of air-drying plant.
- Gradual drifting of set points due to vibration in the system. Keep a continuous watch on the deviation of the controlled parameter from the set point and the tolerance limits. If it is observed that the parameter has a value on both upper and lower sides of the set point and is within tolerance limits, the process is said to be in control. However, if the parameter has a value continuously on one side (either above or below the set point, and slowly approaching the tolerance limit) then it indicates a gradual shift of the controller and corrective action will be required.
- Control instrumentation is getting affected due to corrosive conditions.
- Connecting cables getting affected by heat from nearby units operating at very high temperatures/hot surfaces nearby.

- Compensating cables are getting disconnected from sensors or input junction points of instruments due to vibration, falling of corrosive liquid due to leaks from overhead piping or falling rain water.
- Ingress of dust, chemical vapours, or moisture through damaged covers of junction boxes for connecting cables.
- Pilot tube connection from sensors to the control valve is getting choked frequently due to salt deposits.
- Protective sheaths of thermocouples getting damaged due to direct impact of flames or high-temperature fluids.
- Deposits of salts or dust on sensors/probes reducing their sensitivity.
- Sensors are too near strong magnetic fields.
- Sensors are too near bus bars carrying heavy current (which can induce magnetic fields).
- Vibrations in sensors due to heavy/high-speed machines running nearby.
- Clogged pressure taps.

8.7   Preventive Measures for Process Control

- Properly calibrated smaller capacity day tanks for feeding materials with metering pumps/calibrated feeding devices on delivery side.
- Provision of battery backup for instrumentation.
- Provision of standby power to key equipment through diesel generator sets (of adequate capacity) with automatic start in case of main power supply outage.
- Provision of alternative cooling water arrangements for preventing overheating of process vessels in case of failure of cooling water pump.
- Prevention of heat loss through process units by adequate internal/external insulation by ceramic paper, insulating bricks, magnesia lagging, and aluminium cladding. This is to ensure proper temperature inside process units.
- Provision for (uniform) flow distribution through spargers, nozzles across the cross section of reaction units, filters, acidifiers, etc. for preventing pockets of concentration or channelling through the units.
- Provision of hydraulic motor for rotating the agitator in a clockwise and anticlockwise direction for process vessels handling viscous materials/which tend to solidify.
- Calibration and maintenance of instruments, hardware, and accessories to be done as per instructions from OEM at regular intervals to ensure accuracy of readings and control action.
- Draw independent samples and analyse them to confirm that readings are correct.
- Make arrangements to prevent vibrations of control valves/set points.
- Make provision to prevent drifting away of set point of control systems.
- Design and install systems which are independent of each other for process control.

- Different types of indications for same process parameter for assessing of process control. (Example: temperature and composition of gas streams shall be simultaneously monitored for accurate control). This can be cross checked through feed rate of fuel.

8.8 Limitations of Control Systems

Feed Forward

This can be used when relationship of output parameters with input conditions is well known.

Example: Furnace exit gas temperature is directly related to feed rate of fuel oil. This type of control cannot be exercised if the relationship is not accurately known.

Feedback Control

This can be used for smaller reactor/process unit capacity (volumetric capacity, heat capacity).

The change in value of exit process stream is monitored continuously and compared with the set point (value). The operating conditions like feed rate of reactants are changed as per deviations from the set point. However, this action can continue till another deviation in opposite direction takes place. Thus, the process stream values hover (fluctuate) above and below the set point.

8.8.1 Hovering of Values of Process Parameter

The hovering is more if the volume of process equipment is large since it may take more time for the change to take place. Hence, the controlled parameter will be a little more and a little less alternately than the desired values as per limits provided for control action.

The required value of the parameter thus lies in between these limits with feedback control. It also depends on the accuracy of the sensors and the arrangement (whether electrical or mechanical) for conveying the signal to the controller, and mechanism for control action.

8.8.2 Typical Examples

8.8.2.1 Examples of Feedback Control

Control of pH of Exit Process Stream by Adjusting Flow of Alkali to the Reactor

Alkali addition to the reactor is started as per signal received from the sensor when the pH of the exit process stream falls to a particular value below the required pH.

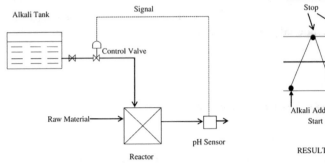

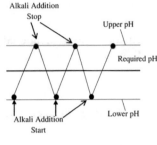

RESULTANT pH IS BETWEEN UPPER
AND LOWER LIMITS

FEED BACK CONTROL OF pH

Fig. 8.1 Feedback control of pH of reactor

This addition does not stop exactly when the required pH is reached, but continues for a little more time till the pH increases to a particular value above the required pH. The alkali addition is stopped when the pH reaches the particular value above the required pH and another signal is received from the sensor. The pH is thus maintained between the lower and upper values of the required pH.

Alkali is thus not added continuously. The rate of addition depends on the volumetric capacity of the reactor and the flow rate of the exit stream (Fig. 8.1).

Control of Air Supply Pressure to Reactor

A compressed air receiver is used to supply compressed air at a certain pressure to a reactor for air oxidation reaction. Air pressure to the reactor is maintained at the desired value by supply of compressed air from air receiver.

Control of pressure in air receiver is done by automatic switching on/off of air compressor generally mounted on the air receiver.

The air compressor is switched on if the pressure in the receiver drops to a lower set point and continues to run till the pressure rises to the upper set point. This is able to maintain the required air pressure in the air receiver for supply of compressed air to user.

The compressor is not run continuously.

Control of Temperature in Reactor

An agitated steam jacketed reactor is to be operated by feeding liquid raw materials. The temperature of the exit stream from the reactor is to be maintained at a particular value. Steam is available constantly from main line to the pressure regulator *before* the jacket. Operation of the steam pressure regulator for supplying steam to the jacket is to be controlled as per deviation from set value of temperature of the exit stream.

Steam supply to the jacket of the reactor is started when temperature of the exit process stream falls to a particular point below the required value and a signal is sent to the steam pressure regulator. The steam supply does not stop exactly when the temperature of the exit process stream reaches the required value; but continues for a little time more till the temperature increases to a particular point (value) above the required temperature and another signal is sent to the steam pressure regulator.

The required temperature is thus maintained between the lower and upper points (values) of temperature. The steam supply to reactor jacket from the pressure regulator is not continuous. It depends on the volumetric capacity of the reactor and the flow rate of the exit stream.

8.8.2.2 Examples of Feed Forward Control

Example 1: Manufacture of Sulphuric Acid The air flow and sulphur feeding are generally well adjusted for a particular production rate of the plant, and are not adjusted repeatedly (which will be the case if a feedback control is provided for the plant).

Rate of sulphur feeding can be directly adjusted by setting on the control valve/sulphur feed pump and air flow is set by speed of the air blower (positive displacement type blower).

It is now desired to increase the production rate of this plant. It will require higher inputs of air and sulphur. This can be done by increasing the setting on the sulphur feed pump/control valve and speed of the air blower since there is a direct known relationship of sulphur input and air flow with the rate of production.

Air flow is increased to facilitate proper burning of the additional sulphur (and to control the $SO_2\%$ accordingly in the gas at inlet to catalytic converter). Air volume delivered by the blower has direct relationship with its speed in case of positive displacement-type blower.

The relationship of these actions on the rate of production as well as system temperatures, additional consumption of water for chemical combination and increase in steam generation by the WHRB is also known. Hence the feed forward control of sulphur input and air flow can be exercised properly.

Example 2 Increasing the cooling water flow to the ammonia condenser of a refrigeration plant for bringing down the discharge pressure of the ammonia compressor is another example of feedforward control.

There is a direct relationship between cooling water flow rate and the discharge pressure of the compressor. The cooling water flow is not repeatedly adjusted as per the discharge pressure of the compressor.

8.8.3   Practical Guidelines for Instruments

A chemical plant can be run safely, without causing environmental pollution, and efficiently when appropriate instruments are provided for monitoring and controlling the process conditions.

It is necessary to regularly check the accuracy of the instruments used for monitoring process conditions such as temperatures, pressures, flow rates, pH, levels in various tanks, analysers of exit process streams from ETP and stacks; and working of safety devices and electrical safety trip interconnections for safe and efficient working of the process plants.

Recalibration of important instruments must be carried out in case of inaccurate indication of process parameters.

General guidelines for this could be as follows:

- Make a list of important process parameters to be monitored. The requirement could be for indication, recording, controlling, or all of these.
- It may serve the purpose better when the instruments indicate both in the field (at the process units) and in the control room also. Hence the instruments should have field sensors, auxiliary contact points, connecting cables, and indicators in control room.
- The maximum expected value of the temperature, pressure, flow rate, etc. (for example: in deg. C, kgs/cm^2, m^3/h) should be mentioned while procuring the instruments.
- The desired accuracy shall be specified *in absolute terms* instead of as a percentage of full-scale reading.
- Properties of materials being handled, piping sizes, desired place of installation, size of indicating dials, frequency of recording parameters, etc. shall be mentioned.

The Manufacturer Should Provide
- Weatherproof enclosures for the instrument, cable junction boxes, etc.
- Test certificates for the instruments after calibration.
- Sufficient number of spares like sensors/probes/protective sheaths while purchasing.
- Specifications for all these spare items.
- At least three copies of instruction manuals (one master copy for head office, one for plant operating personnel and one for instrument engineer). Detailed procedures for installation, calibration, precautions to be taken during operation, and minor repairs should be given in the manuals.
- The purchase order shall include a clause for assistance by the manufacturer for annual overhaul and carrying out necessary repairs free of charge for 2 years from the date of commissioning.

General
- Care should be taken during positioning of the sensors/probes that they are not subjected to a corrosive environment (provide a thermowell/protective sheath of

suitable material as enclosure). Strainers shall be provided in sampling lines to protect the sensors from damage due to hard erosive particles in the fluids.

- The sensors and instruments shall not be subjected to vibrations.
- Route of the connecting cables shall be away from bus-bars carrying heavy currents and very hot surfaces. There shall be no chance of damage due to dripping of corrosive fluids from above due to any leak from the overhead pipelines.
- Instrumentation engineers and technicians should regularly check the system for any damage to sensors/probes/protective sheaths, connecting cables and junction boxes, accuracy of the instruments, etc.
- Proper test kit, standardised chemicals, and facilities for calibration should be available in the plant.

Independent Check

It is advisable as a practical guide to independently check the readings/indications shown by the instruments.

Examples

- Check level of liquid in a tank by dip stick/immersing a rod in the liquid to confirm the reading shown by the level indicator in control room panel. This is especially useful for tanks which are likely to overflow during filling by pumps located at a distance. It is necessary to correctly know the level in order to stop the pump.
- Check the concentration of sulphuric acid by titration against standard NaOH solution; by measuring the rise in temperature when diluted with known amount of water in order to confirm the reading shown by conductivity analyser.
- A remote sensing instrument may indicate only 70 °C as temperature of external surface of the shell of a process unit. If it is doubtful, then one may rub a piece of solid sulphur on the surface. If it melts, then the temperature is about 120 °C. If the piece of sulphur catches fire, then the temperature is not less than 250 °C.
- Experienced engineers can also watch the colour of the flame in a furnace; the presence of soot in the exit gases and correlate with temperature in the furnace and proper functioning of the oil burner.

Plant engineers and operating personnel should also be alert and detect abnormal smell near process units or electric motors, noisy running of agitators or pumps, current drawn by drive motors of agitators/pumps and vibration in process units, etc. to verify whether process parameters are going out of control or some breakdown is likely to take place. However, this needs keen observation and experience.

References/for Further Reading

1. "Process Equipment Procurement in the Chemical and Related Industries", Kiran Golwalkar, Published by Springer International Publishing, Switzerland.
2. "Production Management of Chemical Industries" Kiran Golwalkar, Published by Springer International Publishing, Switzerland.

Chapter 9
Practical Considerations and Guidelines for Project Managers, Plant Managers and Plant Engineers

Abstract All topics mentioned in key words are explained in detail in individual sub-sections of this chapter. Important considerations for establishing and operating a chemical plant, scope of supply and conditions for performance guarantee test to be included in purchase order for a new plant; precautions to be taken for expansion of capacity, revival of old plants, and relocation of a plant are given. Situations which indicate modernisation and diversification of existing plants are explained. A typical format of activities for modernisation shows details of the necessary matters.

Guidelines are given for project manager, plant managers as well as maintenance engineers for smooth erection, commissioning, operation, and maintenance of the plants. Check points are given for operation of effluent treatment and air pollution control facilities.

Many plants generate corrosive, reactive, toxic wastes. Safe disposal of such dangerous wastes is also explained.

Chemical engineers shall coordinate with electrical department regarding normal and maximum electrical load, provision for emergency power; and installation of electrical equipment, etc.

Keywords Statutory permissions needed for establishing and operating a chemical plant · Purchase orders for a new process plant · Check points for relocating a plant · Expansion of capacity of existing plant · Revival of an old/idle plant; decontamination of old units · Modernisation of an existing plant · Situations which indicate need for modernisation and revamping · Study of the past plant performance · Diversification to new products · Guidelines for project manager · Plant manager · Project engineers · Plant engineers and maintenance ngineers · Coordination with electrical engineers · Effluent treatment and air pollution control · Disposal of dangerous wastes · Water treatment plant

© The Author(s), under exclusive license to Springer Nature Switzerland AG 2022
K. R. Golwalkar, R. Kumar, *Practical Guidelines for the Chemical Industry*,
https://doi.org/10.1007/978-3-030-96581-5_9

9.1   Statutory Permissions and Compliances

The Plant manager should have the statutory permissions for the following which should be valid during operation of the plant/up to the last day of operation (when the plant is to be stopped for annual overhaul and maintenance.) These are necessary for proper legal and safe working of the plant units.

9.1.1   Statutory Permissions Needed for Pressure Vessels

- Statutory permissions applicable to all pressure vessels in the plant; boilers, economisers, high pressure feed water lines, steam lines; compressed air receivers; and high pressure units in refrigeration systems.
- Only qualified and licensed boiler attendants shall be engaged to look after the boilers and economisers. Their certificates should be valid while attending duty (for the heat transfer area and at the maximum working pressure) on the plant boilers.
- The settings of safety valves and rupture discs on pressure vessels (and the vessels which may get pressurised during operation) are to be offered for inspection, for setting the release pressure and final approval by statutory inspector. These settings shall be tamper proof (locked).
- Setting of all required safety devices and systems are also to be done as per instructions from statutory authorities.
- Any repair/maintenance work carried out on equipment and pipelines operating under pressure must be offered to statutory inspectors for final testing in their presence and approval.

9.1.2   Statutory Permissions for Storages and Electrical Units

- Storages of process units handling dangerous materials (CRIT corrosive, reactive, inflammable, toxic): These must have dyke walls to collect any spillage. The quantities stored at the premises at any time shall not exceed the limits prescribed by pollution control board authorities.
- Storages of fuels and high pressure gases: These must have cordons all around and provision of firefighting and lightning arresters.
- Permission for power supply connection shall be obtained.
- Permissions are also required for electrical installations such as transformers, rectifiers, HT lines, HT motors, metering panels, electric motors, and lighting (flame proof type). Sump pits should be provided below transformers to collect any oil leaks.
- Proper firefighting facilities should be installed for transformers.

- Copy of certificate for the licensed electrical technicians to work on high tension lines shall be available and shown to electrical inspector when asked for.

9.1.3 Compliances with Statutory Requirements

It is necessary to comply with following directives and instructions for obtaining approvals. Compliance reports shall be submitted to competent authorities for:

- Inspection of boilers
  - Registration of boilers.
  - Issue of provisional orders to operate boilers.
  - Steam pipeline drawing approval.
  - Endorsement of certificates of boiler operation engineers, boiler attendants.
  - Testing the steam generating boilers, economisers, superheaters, and high pressure steam and feed water lines at the recommended test pressure.
  - Setting of safety valves and their setting.
  - Copy of certificate for the licensed boiler attendants.

- Disaster management plan shall be submitted for approval by the authorities.
- Environmental impact assessment (EIA) study shall be carried out and submitted for statutory approval before commencing construction of plant.
- Record of load tests for lifting devices like hoists, EOT cranes, chain pulley blocks shall be up to date.
- Building plan approval.
- Building occupancy certificate.
- Confirm and report proper working of firefighting pumps, installations, and arrangements of portable devices in areas susceptible to fires.
- Structural integrity of all load bearing structures. Regular inspection shall be carried out through licensed experienced structural engineers. Recommendations for repair/reinforcement must be immediately complied with.
- Permission for ground water use, if ground water is the source
- Allocation/permission for use of surface water source.
- Permission under hazardous waste (handling and management) rules.
- Confirm the adequacy of effluent treatment and air pollution control systems for maximum expected load of pollutants which may occur during disturbed working of plant due to any reason. Provide higher capacity equipment if any shortfall is noticed.
- Approval of factory plan.
- No objection certificate from fire department.
- Provision of adequate/multiple escape routes in the plant layout. These shall be kept unobstructed always. Approval shall be obtained from factory inspector.
- Ensure provision of exhaust systems (with scrubbers and collection hoods) for working areas to ensure safety of personnel.

- Special sanitisation sections at entry and exit for all employees.
- Rest room facilities for workers.
- Separate rest rooms and washrooms for women.
- First aid facilities and emergency transport (ambulance).
- Submit details of arrangements which will be made as per safety and pollution control in order to obtain consent to establish the plant before start.
- Install necessary safety devices and pollution control equipment and offer for inspection by competent authorities. This is necessary for obtaining consent to operate the plant as per facilities installed.
- The standard instructions by Directorate of Industrial Safety and Health; by Central Pollution Control Board and any other additional instructions given by state/local authorities must be complied with.
- The plant facilities are generally inspected regularly by factory inspectors, boiler inspectors, electrical inspectors, and pollution control inspectors. They should be satisfied about the conditions of these facilities.
- Make arrangements to comply with (1) disaster management plan (2) rules and regulations set by local authorities.

9.1.4   Additional Requirements

In addition to above, any other instructions from competent authorities for safe working and preventing environmental pollution and fire hazard must be immediately complied with. The facilities for control of fire and release of pollutants and their mitigation must be up to date as required by authorities.

9.2   Purchase Orders for a New Process Plant

9.2.1   Introduction

These shall be drafted by the project department in consultation with marketing department, production engineers, safety department, maintenance engineers and submitted to senior management for approval. The commercial terms and conditions shall also be checked before placing order.

Effect of local seismic and climatic conditions on plant equipment and their lives must be discussed with vendors (as well as designers and fabricators). Suitable reinforcements and corrosion allowances for external corrosion should be provided to all such equipment.

It should be possible to run the plant safely at the given location at prevailing ambient conditions of temperature, rains, humidity, snow fall, seismic conditions, high wind velocities, etc. All equipment should be designed and constructed accordingly.

The following should also be clearly defined in the purchase order.

- Battery limits/scope of supply by seller.
- Obligations of buyer and facilities required by seller at site.
- Vendor must provide ETP and APC to treat the effluents so that the final discharge streams are able to meet statutory requirements.

9.2.2   Purchase Order (PO)

Purchase order should specify:

- The products to be manufactured and specifications for each of the products. Maximum permissible limits of impurities in the products must be specified.
- Rated capacity as MT per day (MTPD) and MT per year (MTPY) for each of the product. This enables working out the annual revenue generation.
- Some tolerance limits may be specified for the rated capacity.
- Running time of the plant per year (generally 8000 hours for a continuous process plant is specified) or.
- Number of batches of products per year for a plant which is run in a batch-wise manner.

Turndown ratio: It should be possible to run the plant a little above rated capacity (generally about 110–115%) or below (65–70%) without loss of efficiency.

- Vendor must clearly inform specifications of inputs (raw materials, water, power, compressed air, steam, etc.) while accepting the purchase order (PO). The vendor of the manufacturing plant must include facilities for pre-treatment/required treatment of the raw materials in case the specified quality of raw materials cannot be procured by the buyer. Vendor may charge additional cost for provision of such facilities after discussing with the purchaser.
- Consumption of raw materials, power, water, steam, fuels, etc. should not exceed certain specified amounts per MT of finished product (as per specifications given already). These should be mentioned in the PO.

9.2.3   Typical Terms and Conditions of Performance Guarantee Test

- A continuous run of 72–168 h or for a certain number of batches.
- Achieving rated production capacity.

- Maximum limits of consumption of raw materials and utilities per unit (MT) of product (as per specifications).
- Vendor should specify the quantities of all waste products (effluents) which may get generated and finally discharged per unit of product along with their analysis. These must be within norms prescribed by statutory authority.

9.2.4   Scope of Supply by Vendor Should Include

- Assembly drawings and materials of construction MOC of all units.
- The mechanical integrity of all the process units and all machinery handling dangerous materials, operating at high temperatures and pressure as offered by vendor must be looked into carefully by technical team of the purchaser.
- Facilities for water treatment, effluent treatment, and air pollution control must also be included in the scope of supply.
- Standby units shall be provided for key equipment and machinery.
- Proper safety devices and interconnections, warning alarms, and detectors for dangerous gases and materials must be provided; and their satisfactory working must be demonstrated by the seller.
- Effect of local seismic and climatic conditions on design and construction of plant equipment must be discussed with vendors (as well as designers and fabricators). Suitable reinforcements and corrosion allowance should be provided to all equipment.
- Detailed Instruction manuals shall be provided for erection, commissioning, operation and maintenance of all key units. Special precautions to be taken during operation and maintenance must be highlighted.
- Supplier must also offer replenishment of catalyst and other additives; spares for process units and machinery either through own sources or recommending vendors who can supply these items of standard quality.
- Quality assurance plan (QAP) must be prescribed and agreement obtained from the suppliers, fabricators and vendors.
- Necessary Instruments should be provided for monitoring and controlling all important process parameters.
- Automatic devices shall have manual override arrangements where necessary.
- Standby pumps and key machinery must be in installed position as far as possible.
- All process units, storages, ducts, and piping which are required to handle corrosive, inflammable, and toxic materials must be designed and fabricated *as if they are pressure vessels* with extra safety margins and appropriate materials of construction MOC.
- Machinery for handling materials should have inbuilt safety alarms and devices, limit switches, and safety interconnections.
- MOC for their fabrication and construction should be selected as per suitability for the materials to be handled. They should be approved by purchaser's engineers.

- Safety devices and interconnections for appropriate units for tripping in case of high pressure or high temperatures should be provided.
- Supplier must demonstrate the satisfactory working of all such safety devices and interconnections before the plant is accepted (purchased) by the buyer.
- Adequate internal thermal protective lining should be provided for units operating at high temperatures. External insulation and cladding on all hot surfaces should be provided to minimise heat loss and to provide protection to personnel in the plant.

HAZID and HAZOP studies should carried out before commissioning the plant and necessary changes shall be done by the supplier in the process units, ducts, and piping if required by the purchaser's engineers (generally at no additional cost to the purchaser).

9.2.5 Commercial Terms

These should include the following conditions -

- Battery limits and scope of supply by seller and obligations of buyer; facilities required by seller at site to be confirmed by both purchaser and seller.
- Technical representative/competent engineer of the vendor should be present during erection and commissioning; should visit every 3 months or whenever required by purchaser; and during first annual overhaul and restart thereafter.
- Penalties for delay in implementation of the project or not meeting performance guarantee terms shall be specified.
- Payment terms and schedule.

9.2.6 Purchasing a New (Major) Equipment

It is imperative that the manufacturing units are run safely, efficiently, and smoothly with minimum interruptions and breakdowns. This needs a detailed study of the past performance of the plant equipment (actual output of marketable grade products, consumptions of all inputs; breakdowns and mishaps which have occurred; expenses incurred on spares and on environmental pollution control) to determine the main causes. This enables including appropriate (corrective modifications) features when purchasing new equipment.

To Be Specified in the Purchase Order
- The required performance/output and guarantees required.
- Battery limits of scope of supply.
- General assembly (GA) drawings with details of internal arrangements.
- Locations and orientations of nozzles for all process streams and utilities.

- Quality assurance plan (QAP) for the equipment (with details of MOC and bought out components like valves, level indicators).
- Safety devices provided.
- List and specifications for spares.
- Maximum safe operation limits (for temperature, pressure, speed, etc.).
- Specifications of utilities available at site. Vendor shall inform maximum expected consumptions of these utilities while using the new equipment.
- Details of external thermal insulation and cladding required.
- Installation and commissioning assistance *required by purchaser.*
- Delivery period/implementation schedule.
- Stage as well as final inspection schedule.
- Site facilities required by vendor (including civil foundations).
- Loading, transportation, and unloading arrangements.
- Supply of erection, commissioning, operation, and maintenance manuals on confirmation of purchase order.
- Training of purchaser's personnel for operating and maintaining the new equipment.

It is advisable to take preventive actions for getting safe and reliable performance in the future by ensuring following checks:

- Quality of raw materials, utilities, and other inputs should be as specified.
- ETP and air pollution control systems as well as disposal arrangements for wastes must be in operation before starting the plant with new equipment.
- Calibration of instruments.
- Proper operation of all control valves and safety devices must be confirmed.
- Proper training for operating and maintenance personnel.

9.2.7   Relocating a Plant

Generally, a manufacturing plant is set up at a site after careful selection of the process, the equipment required, and considerations for the climatic conditions at site. The process units and plant machinery are procured accordingly and erected at site. The plant is then commissioned to start manufacturing of the products. In case of a turn-key project all equipment are generally supplied, erected, and commissioned by a single party. However, this takes considerable time to carry out all such necessary activities; and there can be delays due to practical reasons. Hence in certain cases a decision to buy a running/idle plant from another location (which could be in another country) and to relocate it at the selected site in one's own country may have to be taken.

9.2.8 Typical Criteria for Such Decision

- Market Intelligence (research) has confirmed that there is good demand for the products (within reasonable distance from the selected site); and there is no existing manufacturing unit nearby.
- Considerable money and time can be saved on design, engineering, preparing specifications, and documentation which are generally required while purchasing a new plant/equipment.
- The cost of purchasing an existing plant by dismantling and reconditioning the existing equipment, shipping it to a new location, and reassembling can be significantly lower than purchasing a completely new/turn-key plant which can be much more.
- Obtaining only the technical know-how (Basic and Detailed engineering packages) and building the plant through own efforts may not be possible due to technical or commercial reasons.
- It has become clear that it will take too long to set up the plant with own efforts (even after engaging contractors).
- Some key equipment cannot be procured in a reasonable time in one's own country even after trying from many sources while these are available readily in the original plant.
- It will be convenient to dismantle the old units and erect at the site in lesser time.
- It is reliably learnt that a running/idle plant manufacturing same product at approximately similar capacity is available for sale (may be for a low price) compared to the price of a new plant or that of a plant constructed through own efforts.

9.2.9 Preliminary Assessment

- On getting this information, a visit by a team with senior technical and commercial experts can be undertaken to evaluate the actual conditions and understand further details about the plant which is to be relocated.
- This visit to the plant should confirm that it is running at a reasonable rate with consumption of raw materials and utilities within normal limits or condition of plant equipment is good enough if it is an idle plant. It may take a few days for this confirmation.
- If the plant is operational prior to relocation check the performance of all process units and machinery before shutting it down for dismantling.
- The raw materials and other inputs required for the plant shall be confirmed to be available at the selected site at reasonable cost.

9.2.10   Check Points

- Appoint a technical team with experienced personnel for the job.
- Study the process technology in detail and make a list of all the process units, auxiliary equipment and plant machinery necessarily required for the planned products.
- Carry out HAZID studies for the process and technology of the plant before purchase.
- Check whether all such items are actually available in the plant (being offered for sale) and their physical condition. **Estimate their residual life**.
- Thoroughly check all existing process units, plant machinery, and piping after properly decontaminating and stopping the process equipment at its existing location. This is to ensure smooth operation after relocation at the new site.
- Check the process flow diagrams, P & ID drawings of the existing plant. Photographs of key units can be useful when the plant is reassembled in its new location.
- Assess the condition of each component. Some may have to be repaired, over-hauled or reconditioned, while other parts can require replacement
- Determine thickness of shell of various units by ultrasonic instruments. Examine for any damage which may have occurred during operation earlier; *and any repairs carried out previously.*
- Carry out pressure tests on all pressure vessels and piping.
- Check all existing safety devices and interconnections. Assess whether these will be adequate or some additional devices will be required after relocation.
- Examine thoroughly all process piping and gas ducts for corrosion.
- Examine integrity of all reactors, heat exchangers, condensers, evaporators, etc. by pressure testing. Check heat transfer surfaces.
- Run the blowers, compressors, pumps, EOT, belt and pneumatic conveyors, bucket elevators, refrigeration units, etc. on trial and observe their performance.
- Thoroughly inspect all electrical motors, starters, cables, transformers.
- Look into their specifications for requirement of power input. These should *match* with power supply *available at site* with respect to voltage, phases, frequency etc.
- Check activity of catalyst in the converter and its **residual** life.
- Make a list of all items which can be dismantled as such and transported to the site.
- Make a list of items which can be *only partially* dismantled and some useful materials recovered for use at the new location (units of large size like storage tanks, units with internal lining of refractory/acid-resistant bricks)
- Estimate the cost of dismantling, recovering, packaging old equipment, shipping to selected site, and erecting them again. Compare with the cost of purchasing new equipment *which may be less.* This shall be worked out at as early as possible.

- Make a list of items which may have to be discarded (spent catalyst, damaged brick lining, damaged demisters, belts, buckets, thermocouples, etc.). *Estimate salvage value.*
- Clear any pending payments to local statutory authorities and dues to be paid to work force and any other local stake holders.
- Introduce a clause for performance guarantee for the plant/individual process units and plant machinery at new location if the job is being done *through consultants.* This may not be possible always however.
- Keep ready all necessary civil foundations, derricks, tools, utilities and experienced manpower at new site before the equipment arrive from old location. All such arrangements shall be ready before taking up erection and commissioning activities. This will save considerable costs and delay in project execution.
- All necessary documents and paper work shall be ready to prevent delay in clearance at ports and during local transportation.
- Special precautionary arrangements shall be made at site for safe storage of all process units, instrumentations and key plant machinery, process pumps etc.
- Establish adequate environmental pollution facilities at new site.
- Obtain statutory clearances, permissions, and consents for pressure vessels, electrical installations, pollution control facilities, all safety matters, etc.

9.2.11   *Estimate Expenses Required for*

- Visit to the plant by the team with technical and commercial experts.
- Various tests for assessing the condition of existing equipment in the plant.
- Dismantling, packaging, and shipping.
- Making arrangements at site for erection and commissioning.
- Arranging the required balance items (if all the original units could not be transferred).
- HAZOP studies and subsequent changes to be carried out.
- Obtaining statutory permissions.
- Arranging power, water, manpower, fuels, chemicals for effluent treatment, starting materials, etc.
- Calibration of all instrumentation.
- Arranging for marketing of products *including generation of initial off-spec products.*
- Provision of additional units to suit site conditions such as refrigeration system if volatile materials are to be handled. *Example: Original plant may be located in a place having cold climate, while selected site may be in an area with hot climate.*
- Providing air-cooled units (if there is scarcity of water at site).
- Providing oil–/coal-fired units (if natural gas is not available at site).

9.3   Expansion of Capacity of Existing Plant

Obtain confirmation from market intelligence (research) that there will be future increase in demand for products. Thereafter plan accordingly for increase in production capacity (and the product mix). Sales department should confirm their ability to sell the increased production.

If the demand is seasonal, then it should be estimated as accurately as possible for different seasons.

Check maximum present production capacity available and confirm the increase required to meet the seasonal demand.

9.3.1   Practical Guidelines for Capacity Expansion

Consider the plant as consisting of various sub-systems as below:

- Raw material storage and material handling systems in the plant and in the premises separately.
- Processing/Reaction section
- Structural supports for all units
- Product collection and purification
- Effluent treatment and air pollution control
- Utilities section (electrical units and network, water treatment and distribution, steam generation and use, etc.).
- Maintenance facilities
- Manpower required for all operations in the process plant and premises.
- Civil works and structural supports
- Process piping and ducting
- Instrumentation and control
- Process control laboratory
- Storage of finished products.
- Arrangements for dispatch of finished products.

Check Each Sub-System
- Check each and every unit and machinery in above sub-systems for adequate capacity and mechanical integrity/strength as necessary for the increased production capacity.
- Check all structural supports, load bearing columns, beams, and platforms.
- Make arrangements for storing additional raw materials and their handling.
- Check the total electrical load (running and standby); rating (HP) of individual motors and capacity of power cables. Obtain sanction for higher maximum demand from power supply grid authority if there is going to be an increase in the electrical load.

- Make arrangements for reducing consumption of water. Work out demand for additional water supply, prepare water budget, and obtain sanction from statutory authority.
- Augment capacities of effluent treatment, air pollution control, and waste disposal facilities.
- Make arrangements for recycling maximum amount of treated effluents.
- See that no congestion takes place in the plant nor movement of men and materials is hampered due to additional equipment to be installed.
- Carry out HAZID and HAZOP studies before starting the plant with increased capacity. Install additional safety devices and make necessary changes in systems.
- Provide more instrumentation and warning alarms.
- Increase facilities for laboratory analysis of raw materials, process streams, and finished products.
- Storages of non-compatible materials, CRIT (corrosive, reactive, inflammable, toxic) items, etc. to be arranged properly and insurance cover to be taken.
- Need for additional manpower to be estimated and provision to be made in consultation with human resources department and workers' unions.
- Make arrangements for facilities to be provided for additional manpower.
- Make arrangements for storing additional products and their dispatch.

9.3.2 Augment Existing Process Units and Machinery for Plant Expansion

- Certain units do not need any changes if they already have built in higher capacity, e.g. water pumps, conveyor belts, refrigeration plant, and steam generation boiler.
- Some units may need minor upgrading, e.g. filter press, plate heat exchangers (wherein some more plates can be adequate to meet the additional load).
- Some units may need major changes, e.g. bigger motors for circulation pumps, additional catalyst, additional tower packings after increasing the height of absorber tower, and provision of additional steam coils in evaporators.
- **Replacement of existing units**, e.g. gas to gas heat exchanger, main air blower if the existing equipment are already running at their peak capacity (in case of a sulphuric acid plant), and bigger circulation pumps for acid system.
- **Additional** new process units to suit the new product mix planned, e.g. new oleum tower and circulation systems if it is planned to manufacture oleum in addition to sulphuric acid in the same plant.
- New civil foundations wherever required.
- Check effluent treatment plant capacity and augment individual equipment. Make arrangements for disposal of additional dangerous wastes if they may get generated.

- Check tube settler, filter press, active carbon filter, bag filter.
- Augment the air pollution control facilities, e.g. provide more candle demisters and bigger alkali circulation pumps for scrubbers.
- Approval of these augmented facilities and consent will be required from statutory authorities to operate the plant at increased capacity since more effluents may get generated.

9.3.3   Revival of an Old/Idle Plant at Present Location

One should find out the reasons for the plant to have become idle before attempting to revive it at the same location. Some typical reasons could be the product was not being sold in the market (due to poor quality, or availability of cheaper products in the market), it had become unsafe to operate, raw materials or utilities were not available as per requirement, major breakdown of key equipment, etc.

The revival may be attempted if the management is confident of overcoming the above problems; the products can be sold through marketing network available or used as intermediates in the running plants with the organisation.

9.3.3.1   Decontaminate

Carry out a visit to the plant for a careful inspection of the existing equipment and facilities to assess their actual condition and capacity. It is necessary to carefully decontaminate the units before a proper inspection. It needs a team of experts and certain equipment for these operations which could be costly as safety issues will also be involved.

They will involve removal of residual materials (some of which could be toxic, inflammable, or corrosive) from various process units like reactors, condensers, and circulation tanks. These materials can be carefully stored for future use or it may be required to neutralise such materials and safely disposed of.

Firefighting facilities, exhaust hood with high capacity exhaust fans, scrubbing system, storage drums, etc. will have to be kept ready for the decontamination operations.

Dilute solution of alkali and circulation system, flushing water shall also be available. Arrangements should be made to treat the additional effluents generated during the cleaning and decontamination operations.

Individual equipment should be cleaned thoroughly and inspected (preferably pressure tested). They should be repaired thereafter and tested again.

9.3.3.2 Servicing and Repairs

All mechanical systems and process units shall be cleaned, lubricated, and again run on trial to confirm satisfactory working. Any faults detected shall be rectified and trials taken again.

Typically, the following equipment need thorough servicing and replacement/ repairs to some of the components:

- Material handling systems like belt and screw conveyors, overhead cranes
- All pumps and blowers
- Electrical installations, motors, cables, starters
- Diesel generator sets for emergency power supply
- Firefighting arrangements
- Storage tanks and facilities for storing raw materials and finished products.
- Structural supports, work platforms, load bearing columns, and beams
- Filter press may need some plates and filter cloth.
- Catalytic converter—catalyst bed support and replacement of catalyst (if inactive)
- Absorption tower—internal packings, liquid distributor, and demisters.
- Agitated reactors—agitator blades, shaft, shaft seals, driver motor, and gear box.
- Heat exchangers—cleaning and pressure testing of shell and tubes.
- Water treatment plant and storages for treated water.
- Effluent treatment ETP and air pollution control (APC) systems.
- Units operating at high temperatures—inspect internal refractory lining and external insulation; repair where necessary.
- Instruments—thermo-wells, pressure gauges, thermocouples, sensors for temperature, pH, all control valves, and operating systems.

Cost of calibration and repairs/replacement of all these items should be estimated.

- All pipelines and duct carrying dangerous fluids shall be pressure tested and leaking or weak portions should be replaced. Structural supports for all piping and ducts.
- All valves in the plant (for process units, storage tanks, operating plant machineries).
- Carry out HAZID and HAZOP to confirm it will be safe to revive the plant.
- Provision of additional apparatus for necessary laboratory work.

9.3.3.3 General

Fresh permissions are to be obtained from competent statutory authorities for all pressure vessels, boilers, and economisers; fuel storage and handling systems; Effluent treatment and air pollution facilities after they have been attended and made ready for use.

Consent to operate shall be obtained before restart of the plant.

Any major changes in the plant layout will need statutory approvals.

Arrangements for initial treatment/purification of raw materials shall be installed if they were missing in the plant. This is necessary to improve the quality of the products.

The existing manpower may be offered employment to continue to work in the same plant (with additional training to operate and maintain). New manpower may have to be engaged if there is need for specially trained manpower due to some major changes carried out during the revival. An agreement with the unions may be necessary. This will be a major decision by the management.

The total cost of revival of the plant as well as the time required shall be worked out again.

Marketing arrangements shall be revised and new customers shall be contacted. This is necessary when the revival is done with aims of expansion of capacity and diversification to **new products**.

The return on investment (ROI) shall be estimated again before investing large funds. Only a basic engineering package shall be prepared initially to minimise the costs. This should be examined in detail by senior technical, commercial and management personnel. Further investment can be done if the return on investment is expected to be satisfactory and loans are sanctioned by financial institutions. Approval from majority of shareholders and board of directors shall be obtained.

9.4   Modernisation of an Existing Plant

9.4.1   Situations Which Indicate Need for Modernisation

The managements shall regularly monitor the following conditions in the production plants. These conditions may develop gradually over some time and their impact on profits may not be noticed immediately if the products have a large profit margin or have a large share of the market). However, these conditions can erode the profits considerably in the long run.

Complacency may even set in the organisation and it may become difficult later on to compete with new manufacturing plants or in the international markets. Hence such conditions which indicate unsatisfactory working of the plants should be addressed by modernisation of the plant.

Check if:

- Cost of production has increased.
- There are complaints from buyers about quality of products.
- Buyers are not giving repeat/more orders.
- Cheaper and/or better products are available in the market.
- It is becoming increasingly difficult to meet delivery schedules; or to supply in lots or packages as desired by buyers.

- Consumption of raw materials and other inputs per unit of product (output) has increased considerably. Consumption of spares for process units and plant machinery has crossed reasonable limits even when the plant is not very old.

*Generally, a little reduction of process efficiency and increased wear and tear of machinery can occur when the plant has been in operation for a few years for which due consideration shall be given. Plant engineers shall provide a **calculated allowance** for such little reduction in efficiency of process; or increased wear and tear of individual machinery on an annual basis. An immediate modernisation with major changes may not be necessary in such cases.*

- Inefficient operation or higher consumptions of spares can occur only when the plant has become very old. This should be controlled well in time.
- Certain equipment and machinery are consuming more utilities like power, water, and steam per unit of output from them.
- Plant has become unsafe to operate.
- Generation of effluents has increased (which has resulted in environmental pollution as well as loss of products/various input materials).
- Plant stoppages have increased due to frequent need for repairs (cost of maintenance has increased).

9.4.2 Situations Which Indicate Need for Revamping

Revamping of the process plant and accessories may become necessary when the performance has got deteriorated considerably (high cost of production, unsafe working, excessive generation of pollutants) and when life of the equipment is reaching their (useful) ends. This is to be particularly looked into the case of pressure vessels, process units, machinery, and piping which are used to handle dangerous, toxic, inflammable, and corrosive materials. **Revamping shall generally aim to improve the safe working, increase the production rate in order to reduce cost of production, get a better share of the market, and a higher return on the investment incurred.**

The amounts that may be available by recovery/sale of scrap should not be considered while estimating the costs.

9.4.3 Study of the Past Plant Performance

The above situations call for improvement in order to maintain profitability while working safely and without causing environmental pollution. A detailed study of the past performance of the plant should be undertaken to arrive at the main causes of deterioration; and to decide corrective actions. This may be done by selecting a period of the immediate last 3 years for the study. It will be very useful if the plant was operating at rated capacity or near about it during these 3 years. The study should carefully consider the following:

Plant Run Time
- Total time for which the plant was running in each year.
- Longest continuous run time between stoppages i.e., longest smooth non-stop working time.
- Time lost due to unscheduled stoppages resulting from breakdowns of equipment and due to power outages (separate data will be required).
- Plant stoppages due to preventive maintenance and scheduled overhauls.
- The data should be compared with standard norms/industry standards of run times and idle times.

Production Rates
- Rate of production (average production achieved per day in the year; and maximum per day during a steady run of at least 7 days)
- Compare with the normal rated capacity, with maximum permissible capacity, and lowest rate operable during steady runs.

Quality of Products
- Check when the quality of products was as per specifications.
- Generation of products which did not meet quality specifications/which were not of marketable grade.

Expenses Incurred on Consumption of Process Inputs and Spares for Maintenance
- All inputs (raw materials, power, fuels, water, steam, etc.).
- Spares for process (catalyst, filter aids and cloth, chemicals for water treatment, absorber tower internals).
- Spares for maintenance (items consumed during normal wear and tear).
- Spares procured from original equipment manufacturers.
- Expenses on reconditioning of spares and *resulting performance thereafter.*
- Repairs done for key equipment (separately record the expenses for minor and major repairs).
- Replacement of *an entire process unit (reactor, condenser) or entire machine (pump, compressor).*

Mishaps and Their Frequencies
- Severe incidents (resulting in major injuries to or loss of personnel; loss of production and raw materials or plant assets).
- Incidents of near-miss when there was no loss (but there was a distinct possibility).
- Incidents of major fires with considerable loss of manpower, production or plant assets.
- Occurrence of minor fires (when there was no loss of manpower, production or plant assets).

(*Determine the root causes of each accident and any near miss that might have occurred.*)

9.4.4 Cost of Controlling Environmental Pollution

- Look into generation of all wastes during normal (steady) run of the process plant.
- Loss of finished products and un-reacted raw materials through wastes.
- Associated costs of recovery from the waste if *it was possible to do so.*
- Treatment of liquid effluents and air pollution control (maintenance of facilities, consumption of neutralisation chemicals, power for operation of these facilities).
- Disposal of solid wastes (either through incineration by own facility/disposal through authorised agency).
- Cost for controlling sudden generation of excessive quantities of pollutants (study the severity and frequency of such incidents) due to process upsets
- *Determine the root causes of each of these incidents.*

9.4.5 Study of Undesirable Incidents

All senior personnel (engineers, technicians, and senior management) should look in to detailed reports of the incidents which have taken place in the past 3 years which were undesirable but were controllable (*and hence should not have occurred*). This study will enable them to take preventive actions. Some of these incidents could have occurred due to:

- Careless operation exceeding the maximum permissible limit (as per design or as recommended by the OEM) of temperature, flow rate, pressure, and speed of rotation during operations. *Study the frequency of exceeding the permissible limits and the duration for which such incidents occurred.*
- Exceeding the maximum power demand beyond the limit permitted by grid authorities.
- Mismanagement of water budget (excess draw from resources) and excess consumptions due to wastage, misuse of treated water; inability to recycle/reuse treated and recovered water from ETP; excess discharge of effluents from premises.
- Not taking sufficient protective steps for process units and machinery during abnormal weather conditions (*which could have affected the plant performance*).
- Interruption in supply of raw material due to delay in procurement.
- Delay in attending to labor problems.
- Mishaps which occurred due to negligence.

9.4.6   Comparisons of Plant Performance

The figures of (1) plant run time, (2) production, (3) expenses incurred on consumptions and maintenance, (4) mishaps (incidents of accidents and their frequencies (5) costs for controlling the environmental pollution are major indicators of **plant performance.**

These are to be compared with the best reported by industrial units producing similar products. Reasons for the difference between own figures and the above should be carefully determined.

The production and maintenance engineers, senior management, stores controller, and marketing department have to carry out detailed analysis of the data for past 3 years as it can highlight the issues and situations which need to be attended on priority for improvement. Some of these are the present operating condition of each equipment available; and assess their capacities for meeting the expected load (after the corrective actions considered) in order to achieve the aims as planned. Compare with design values and maximum permissible limits as recommended by OEM. However, the safe limits prescribed by statutory authorities should never be exceeded.

To look in to the likely reasons for the present unsafe or inefficient situation (which could be overloading of process units, occasionally exceeding pressure or temperature limits which may develop unsafe conditions during running of the process plants, bottle necks in material handling, insufficient cooling capacities, generation of pollutants, etc.)

Determine the root causes after a detailed analysis of each of these.

9.4.7   Important Aims of the Organisation

The important aims of the organisation are to be achieved while modernising the production plant while following the framework of the organisation policy and priorities. This should be done **without violating any statutory rules and regulations**. The corrective actions should be generally designed and carried out to meet the following aims of the organisation:

- Improving safety during operations and maintenance (zero mishap).
- Preventing environmental pollution (zero liquid discharge).
- Client satisfaction through better quality and timely supply of products.
- Reducing cost of production (less consumptions of inputs, better heat recovery, efficient material handling, and improving overall efficiency).
- Modernising the plant and related facilities (better product quality, less breakdowns).
- Expansion of production capacity (if considered feasible after market survey).
- Diversification to more products (to widen presence in the market).

9.4.8 Differences in Assessments

There can be difference of opinion among the concerned personnel while arriving at the likely reasons for the mishaps, inefficient operations, higher cost of production, higher consumptions of spares, generation of excess effluents, etc. Hence many options could be suggested by them for the corrective actions to be carried out for meeting the above aims.

The priorities for corrective actions are to be decided by study of the past performance of the plant, the situations which indicate modernisation on the basis of a **cost-benefit analysis**. The priorities could be revised by senior management in consultation with plant engineers.

9.4.9 General

Consider carefully the following before finally approving any change for implementation:

- The original process designs, operating instructions for starting and stopping the plant, process conditions to be maintained, and qualities of raw materials as specified by plant technology designers.
- The limitations of key equipment (maximum operating pressure, temperature, handling of corrosive materials, etc.).
- Safety devices (pressure relief valves, rupture discs), safety trip interconnections provided.
- **The maximum safe limits for operation recommended by the original equipment manufacturer/permitted by statutory authorities (whichever *is lower*) must never be exceeded.**
- **Carry out HAZID and HAZOP.**
- Technology consultants, technical personnel from original equipment manufacturer OEM and statutory authorities should be informed and their advice taken before introducing changes in the plant.
- Confirm that all personnel responsible for plant operation and maintenance; safety and pollution control; stores inventory control and marketing and administration have fully understood the objectives and methodology for the changes.
- Designate responsible personnel for implementation.
- Estimate costs for each change, the time required, and the outcome (benefits) expected. Senior management shall sanction the funds after a cost-benefit analysis is done.

9.4.10   Modification of the Production Plant and Infrastructure

The corrective actions may be planned and executed by dividing the existing plant systems under the following **typical heads**.

- Raw materials procurement, safe storages, and handling.
- Feeding of various inputs to process units (in controlled quantities by conveying machinery from main storages and storing in day tanks located near process reactors).
- Process reaction units with adequate residence time for reactants, proper MOC/ protective linings, and optimised internals; catalyst loading, suitable arrangements for temperature control, and other accessories.
- Instrumentation for process control and ensuring proper quality of products.
- Equipment.
- Checking adequacy of safety devices, audio-visual alarms, and trip interconnections for all important systems.
- Provision of additional leak detectors and firefighting systems.
- Main transformers and metering panels, electrical motors, starters, cables, busbars, DG sets for standby power, and plant lighting installations.
- Minimising generation and treatment of all wastes.
- Monitoring arrangements for supply of all utilities (water, power, steam, refrigeration etc.) to plant units and improving them where necessary.
- Energy recovery from high temperature or high pressure process streams; and planning for co-generation of power and steam.
- Dispatch of products (proper safe storages, sampling, weighing, bagging, packaging; systems of metering tanks; safe transportation to points of use; and timely attention to complaints of buyers).
- Maintenance activities for process units and machinery, electrical systems, and civil works; keeping standby units and machinery in ready-to-use condition.
- Inventory control of all important inputs and spares (to ensure safe, pollution-free, smooth plant operation for quality products).
- Manpower management (of own personnel), design of incentive schemes for better performance; arranging refresher courses; and taking assistance from external agencies if required in special circumstances).
- Research and development for increasing yields, better designs of equipment, better pollution control, use of cheaper materials, modification of present operations for more safety, and development of new products.
- Innovation of existing facilities for reducing plant idle time, costs of production, and maintenance.

More heads can be added to the above list if considered necessary by the concerned. The programme for implementing the corrective activities can be then finally approved by a detailed technical and cost-benefit analysis. Opinion of senior production and maintenance engineers shall be considered by the management before granting final approval.

9.4.11 Introducing Changes

- Initially minor (and later on major) changes in operating conditions and maintenance procedures shall be introduced.
- Provision of suitable treatment/purification facilities for improving raw materials and other inputs up to required specifications (as given by production and maintenance engineers).
- Innovation/minor changes for upgrading of process units and machinery.
- Major changes in design or construction of process units and machinery.
- Provision of additional process units, energy recovery systems, new machinery (augmentation of plant facilities).
- Change in some specifications of input materials or use of some other (cheaper or locally available) material itself by suitable pre-treatment.
- To work out the various changes/corrective actions required (augmentation of some machinery, modification in designs or replacing some of the inefficient units which cannot be improved upon to the desired extent).
- Costs shall be estimated for each of the changes being considered for the expected improvement in performance. Also carefully consider the time required for implementation.
- The requirement of funds should be carefully estimated after considering various options such as (1) changes in operating procedures, (2) longer production cycles, (3) better maintenance and controlled inventory, (4) standardise spares (5), reconditioning of spares, (6) simplify and strengthen key components of machines, (7) substitute by simpler machines less prone to breakdowns, and (8) reduction in manpower.
- It is necessary to carry out a cost-benefit analysis.
- However, use of a modification to existing process will need considerable changes in the existing plant and may need changes in the designs for the facilities.
- Similarly, use of a new process will need drastic changes in the plant.
- **Incorporation of a modified/new process will be a major decision to be taken by the senior engineers and management of the organisation.**
- These are to be discussed among senior plant personnel, representatives of concerned departments, and management *in view of the past performance, aims of the organisation and priorities (various heads for improvement)*.

9.4.12 Implementation and Monitoring

- It will be convenient and practical to introduce the changes in a phased (gradual) manner. Any major changes shall not be carried out suddenly as they may create safety issues.

- Effect of a change shall be watched carefully and compared with the objectives for introducing it. Stabilise the plant operations, product quality and maintenance after making a change (before introducing the next change).
- Confirm that all the process units and plant machineries are working safely.
- Confirm that facilities for pollution control are also working satisfactorily.
- Should not create operational inconvenience or labour unrest.

9.4.13   A Typical Practical Situation: Modernisation of the Existing Chemical Plant

It is imperative to arrive at realistic cost estimates as well as the time required for carrying out the project of modernisation. The **ROI** is to be worked out before deciding to invest. This type of project needs considerable planning based on market survey and a proper technical assessment of capabilities of existing process units, machinery, and plant facilities. *This is to be done before installation of new process units and machinery.*

All technical details and necessary activities should be carefully considered for arriving at realistic cost estimates as well as the time required for carrying out the Project. It will reduce the requirement of fresh funds if part of existing infrastructure and utilities can be used **as such** or with very minor changes only.

In case this is not possible suitable augmentation of existing infrastructure and utilities will become necessary. Creation of **new** facilities for manufacture of new products should be done if the augmentation is not sufficient to meet aims of the organisation as stated above.

This type of project may be simpler as compared to a new grass root plant since it can be addition of only a few units to the existing facility. However, in certain cases some of the existing equipment may have to be replaced entirely.

The cost of dismantling can be partly recovered by selling the old equipment or scrap generated.

9.4.14   Minimising the Investment: By Examination of Facilities in the Following Order

- Modifications required in existing plant (and use of existing units if possible).
- New additions required for modernisation.
- Modifications that can be carried out while existing plant is running.
- Modifications that can be carried out only when the plant is not running (for a short period) or will have to be stopped for a longer duration.
- **Revamping**: This may involve more replacements of equipment, machinery, piping, etc., and the costs can be on higher side. This should be looked into carefully while filling up the format given below. **Cost-benefit** analysis shall be worked out (Table 9.1).

9.4.15 Diversification to New Products

Some typical reasons when the management may consider diversifications to other product(s) are:

- Demand for current product is reduced.
- Better or cheaper products have become available in the market.
- Cost of production of current product has increased due to high costs of certain inputs.
- The current product is consuming more raw materials.
- The conversion efficiency of the current process is low and generation of effluents has increased.
- Some major/important process unit is nearing its service life and it may not be possible to continue the manufacture of the current products.
- The existing facilities like boiler, cooling tower, refrigeration system, material storing, and handling arrangements have spare capacity and can be used for new products also.
- Sales department has confirmed demand for some other products which are related to or are similar to current products.
- It is desired to increase the profits/market share of the organisation.
- It will be possible to manufacture different products by using the existing facilities as such or by addition of a few units only.
- The organisation set-up is to be kept running even after sales of the main/initial product has considerably reduced. The investment in existing facilities is not to be allowed to become idle.
- It may become possible to use the existing workforce (if it is occasionally idle or partially under-employed) for manufacture of the new products with some training for the new process units and machinery. This will utilise existing workforce fully. Only a few new additional personnel may be employed (only if necessary) for the diversification.

Examples
- Production of 25% and 65% oleums in a sulphuric acid plant by adding oleum towers and SO_3 generation equipment.
- Production of different types of viscose rayon fibre by some little changes in spinning machines.
- Production of HCl in a caustic soda plant by adding facilities for HCl production.
- Production of different types of phosphatic fertilizers from same manufacturing plant by introducing a few additional units.

In certain cases almost the same raw materials and utilities are required for the new products.

A detailed list shall be made for the equipment required for the different products. It will be noticed that many of the existing facilities and equipment can be used as such.

Table 9.1 Format for Modernisation of a Chemical plant

Sr. No.	Major groups of activities required for the Modernisation	Responsibility of	Activity detail	Changes required for Modernisation	Estimate Cost and time required	Desired Product Mix/output
1	**Project initiation** • Perform feasibility study • Establish terms of reference for consultants (supply of know-how) and maximum limits of consumption of raw materials and utilities • Establish a project office	**Project sponsor/stake holders/ investors** **Senior management**	• Examine opportunities and tangible benefits by conducting market survey to confirm demand for the **products** • Analyse cost and output/benefits of various new products • Assess technical and commercial feasibility (expected profitability *after modernisation*) **after** major changes	Examine basic engineering package from consultant; or actions suggested by own personnel for major changes required	Initial cost estimate by consultant	Better product quality and better working of plant without loss of rated output capacity
2	**Detailed project planning** • Establish marketing network and confirm demand for products • Arrange disposal of off-spec products • Set target date for completion (define improved plant working at rated capacities, marketable grade product qualities) • Consider all constraints and prepare budget for project including risk mitigation plan • Estimate required funds for augmentation or dismantling of old units and procurement of new process units and machinery • Plan special resources required (experts/external agencies)	**Project head (director)**	• List out resources required for each job • Identify potential suppliers, vendors, and contractors for process units and machinery • Identify the risks involved • Prepare risk mitigation plan • Products may be off-spec/not of marketable grade during initial runs of plant (**make provision for their safe disposal**) • Identify *local facilities* required at project site	Suppliers of process units and machinery shall be experienced and reliable		

#	Activity	Responsible			
3	**Setting up a project team** • Functional (site) manager • Technical assistants • Logistics assistants • Commercial assistants • Other manpower • External assistance or contractors as per need	**Project head (project director)**		• To make a detailed job list and to establish site infrastructure as per need • Examine present condition of existing units; and assess if any modification is required • Examine requirement of new units, if necessary • To assist for successful erection and commissioning the process units and machinery *within budgeted cost and defined time frame to meet the expected performance after modernisation* • Ensure safety in all activities	Make a list of hook-ups required for integration with existing plant facilities *wherever possible/or need for new facilities*	Prepare a time schedule for erection, commissioning of new units and hooking up
4	**Project execution** • Study P & ID • Execute activities as per planning • Prepare schedule for erection and commissioning, (for attaining deliverables) • Allocate resources for each phase of work • Monitor progress of work phases • Take corrective actions in case of delay	Functional (site) manager		• Plan all activities as per priorities and estimated durations • Prepare a detailed project schedule with BAR charts and CPM technique (using software like primavera) • Plan phases of work (major milestones) • Monitor progress and compare with planned schedules • Ensure safety in all activities	Carefully plan and make arrangements for all activities required for hooking up Emergency plans to deal with any sudden stoppage/problem in existing units must be ready in advance	

5	**Equipment inspection of key units** **Typical static units** • Tanks and furnaces • Heat exchangers • Process vessels and reactors • Distillation and absorption columns • Scrubbers • Pressure vessels **Material handling units** • Pumps • Compressors • Blowers • Mixers • Belt conveyors • Pneumatic conveyors • Hoists and cranes • Screw feeders • Bucket elevators	Site engineers and maintenance engineers reporting to functional manager	Confirm all required process units, machinery and items have been supplied by vendors as required for the new product(s) Follow up in case of short supplies Prepare a detailed check list for each equipment as per OEM recommendation Prepare repair procedure and get it approved from senior engineers Post repair testing as per procedure and preservation till commissioning Check adequacy for pollution control Carry out pressure tests to meet statutory regulations Check all rotary equipment initially for smooth movement without loading Check operation by gradually increased load till rated capacity is reached (if possible) Attend immediately any faults noticed Complete overhaul for big machines like pumps, crushers, compressors, conveyors, and screw feeders. Charge lube oil on compressors and check lube oil circulation	Check delivery of items against project schedule; take corrective actions in case of delay

| 6 | **Process accessories**
• Ducts and piping
• Gaskets and fasteners
• Strainers and filters
• Flow indicator glasses
• Additional valves required for safe operation
• Structural supports for additional loads, if any
• Thermal insulation | Site engineers and maintenance engineers reporting to functional manager | Examine P & ID for new equipment for hooking up with existing ducts, piping, etc. carry out Hazid and Hazop studies
Check all existing support structures also. Reinforce if required
Flush, clean, and check all pipelines by pressure test. Replace weak portions and repeat pressure tests
Conduct radiography tests for pressure vessels and critical joints of process units and piping
Measure thickness of process vessels carrying corrosive, inflammable materials (acids, fuels, etc.)
Provide thermal insulation wherever required
Check all safety valves for their calibration, setting, and mounting. The connecting nozzles must not be clogged
Mark completed jobs on P&ID | | Complete all these jobs before trial runs of the plant |

#	Item	Responsibility	Checks	Remarks
7	**Electrical Installations and equipment**	Electrical engineer reporting to functional manager	• Check all electrical installations and loads on electric motors, transformers. cables etc. Check plant lighting, metering panels, safety interlocks, earth connections,overload trip systems etc. • Check motor control centres, starters and overload protection relays for motors • Check all field mounted push buttons for operation • Check windings, earth resistance, junction boxes, speed, frame size, and greasing for bearings of electrical motors • Confirm power will be available for maximum expected load	Apply for additional power required for bigger motors if required for new units
8	**Provision of instrumentation** as per planning for important process units, process streams and control facilities for manual/automatic control	Instrument engineer reporting to functional manager	Check the following: • All instruments for correct calibration and mounting, safety trips and audio-visual alarm settings. Provision of gas detectors • All control valves for correct operations • Signals to and from control room and field • Correct functioning of emergency switches • Need for additional instruments	

No.	Item	Responsibility	Task	Remarks
9.	**Processing (main reactions and operations) section** • Reactors, mixers evaporators • Condensers, filters • Augment old or provide new equipment required • Reinforcement of structural supports for all additional loads on process units	Process and maintenance engineers reporting to functional manager	Check internals for all units (catalyst supports and activity), filter cloth/ filter media, heating and cooling arrangements, tower packing, demisters, pressure testing of condensers, cooling medium circulation, etc. for safe and smooth working Charge catalyst in the reactors, clean tower packing, demister pads, and fiber bed mist eliminators, clean heat exchangers and pressure test them Get all supports inspected through licensed experienced structural engineers. Provide additional supports if required	Procure new active catalyst well in time before trial runs of the plant
10.	**Product collection and purification** • Purification facilities • Storage tanks and dyke walls system • Sampling systems and dispatch arrangements	Process and maintenance engineers reporting to functional manager	Check distillation units, condensers, vents, heating and cooling arrangements, weighing bridges, filling systems for carbouys and tankers	Clean sludge from storage tanks, attend level indicators, earth connections, vents, etc

No.	Item	Responsibility	Details	Action
11	**Utilities** • Raw water (treated water, filtered, soft, and demineralised water) • Steam generation and condensate collection facilities • Fuel storages and handling • Other heating facilities • Refrigeration and cooling system • Compressed air system • Electrical-power system 1. Incoming power 2. Internal distribution 3. Emergency power 4. Own generation 5. Export of excess power if generated by co-generation units		• Estimate demand for all utilities and fuels after plant modernisation • Clean all storages, attend water pumps, water softeners, ION exchange units, RO and DM water plants. Augment capacities as per increased demand • Fuel storage tanks and handling systems • Clean cooling tower basins, attend water pumps, ID fans • Overhaul diesel generator sets and auto-start arrangements • Attend air compressor, receivers, pneumatic valves, and controls	
12	**Examine civil works** • Foundations and support beams • Building and sheds • Stores • Internal roads • Storm water drains		Reinforce for additional load expected due to operation of existing and new equipment. Repair to seal all seepages/leaks Provide new foundations if required **Check new plant layout and working space available in plant.** Follow instructions from factory inspector	
13	**Check pollution control units** • Effluent treatment units • Air pollution control units • Dangerous or solid waste storage and disposal arrangements	Process and maintenance engineers reporting to functional manager	Comply with statutory rules and regulations Check capacity of individual equipment, circulation pumps, pH controllers, sampling systems, neutralising chemicals	Augment capacities as per need. Make provision for additional funds

#	Item	Responsible	Actions	Remarks
14	**Safety equipment and firefighting systems** Check availability in sufficient numbers and quantities	Process safety officer and his team reporting to plant manager and senior management	Personal protective equipment (portable breathing apparatus, corrosion- and fire-resistant dress, aprons, fall arresters, gas masks and canisters, full length gloves Field persons must be trained to use Flush firewater network to remove all corrosion debris Ensure fire water pumps and pressure switches are in working condition and develop sufficient pressure Ensure deluge system is as per design, all spray nozzles are clean and system will work as per design All hydrant boxes shall be equipped with hose reels and spray nozzles Conduct planned mock drills and get feedback for improvement Conduct evacuation drills for various scenarios such as a major gas leak, fire in plant Conduct surprise mock drill to review the response. Check through external observers to get report for improvements Periodic safety audits to be conducted to ensure safety systems are working properly	Carry out HAZID, HAZOP All important scenarios to be checked and remedial actions to be implemented through standard **Management of Safety Organisation** for the change in process
15	**Additional manpower** • Trained and qualified • Semi-skilled • Unskilled/contractual manpower	Functional (site) manager	Arrange training to operate and maintain modified/new units	

		Senior management		
16.	**Facilities for R&D** • Improve safety • Innovation and development of new products • Reduce costs	Functional (site) manager	Develop designs for better conversion, safer working of plant, use of cheaper raw materials, lower consumption of utilities, new process, and new products	Allot special funds for **R&D activities.** **Check progress regularly**
17.	**Availability of maintenance facilities in plant**	Senior maintenance engineers	Augment existing facilities for efficient maintenance; provide more tools, jigs as per advice from OEM	
18.	**Stores inventory control**	Purchase/ procurement manager	Simplify designs, standardise spares, strengthen machinery, and substitute by easy to maintain units. Check consumption patterns and optimise inventory. Provide standard codes for all items Initiate procurement actions at the earliest for long delivery items and important inputs to maintain the planned start-up schedule Adhere to safety rules, statutory regulations and standard practices for storages of dangerous, inflammable, corrosive items Procure sufficient extra mechanical consumables for erection and commissioning Ensure sufficient couplings, shaft seals and other spares for rotary equipment	It will be better if spares for old units can be used for new units also. This can reduce cost of inventory Avoid too many varieties of spares

No.	Item			Remarks
19.	**Lab facilities to check** Raw materials and finished products All important process streams All wastes generated and exit streams from the plant		Augment existing sampling and testing facilities for getting results of analysis as early as possible for better process and pollution control	May need some additional apparatus and manpower
20	**General administration at site** Human resources, welfare, transport, marketing, liaison with statutory authorities	Functional (site) manager to be assisted by respective staff	**Senior management** Identify (and finalise) external experts, representatives of Process licensors and vendors Verify their credentials/expertise and get approved by senior management Arrange their travel documents and services well before the scheduled date of start-up in order to avoid last-minute hassles Arrange special tools and site facilities required by vendors in advance	
21	**Commercial matters** Finance, taxation, salaries and remunerations, insurance	To be looked after by respective assistants	**Senior management**	Try to use existing facilities after modernisation also

Example

Manufacture of ferric and papermaker's alum, superphosphate require sulphuric acid and treated water. Hence production units for these chemicals can be added to a sulphuric acid plant when diversification is considered.

General

- It will be better to install the additional process units in separate areas of the same premises for operational convenience and to avoid mix-up or confusion.
- Movement of materials or any of the operations in the existing manufacturing set up should not be disturbed when additional units are installed for manufacture of new products. Plant layout shall not become congested when new units are added.
- Existing facilities for supply of steam, power, treated and chilled water, material handling, cooling towers and refrigeration units, and fuel storages as well as handling shall not become overloaded.
- However, facilities for maintenance of process units and plant machinery may have to be augmented if many new units are added.
- Isolation valves shall be provided for supply lines from existing facilities to new units.
- It is advisable that the above existing facilities are run a little below their maximum capacity (at about 80—85%) even when all the products are being manufactured.
- Staggering of running times of equipment drawing more power shall be planned so that they are not run simultaneously. This will avoid overloading of electrical installations. However, sanction shall be obtained from external power supply grid authorities for the maximum power draw when all units are run simultaneously.
- Capacity of effluent treatment and air pollution control shall be carefully checked to handle the additional load. These shall be augmented if more effluents are likely to be generated.
- Facilities for separately monitoring the consumptions of raw materials and utilities in each diversified unit should be installed.
- Process control laboratory shall be provided with additional apparatus and trained manpower for increased analytical work.
- Separate storage tanks and dispatch facilities will be required for different products.
- First aid facilities, safety showers, and firefighting facilities shall be provided separately to each plant (*which manufactures new product*) for immediate availability.

9.4.16   General Considerations

It is a standard practice to carry out **pre-startup safety review (PSSR)**. It generally consists of a complete safety review done by a team of senior representatives from design, engineering, safety, operations, and maintenance departments before starting the plant (*when some major changes have been carried out or new units have been provided*).

Typically, it is an exhaustive checklist to cover the following:

- Process units and plant machinery and support structures.
- The process design and technology should be checked by carrying out HAZID and HAZOP and confirmed to be safe/having provision of sufficient safety devices on all process units and machinery handling dangerous materials; and those which shall operate at high temperature, high pressure, etc.
- Chances of sudden rise of pressure, temperatures or release of toxic, inflammable materials from the process units due to runaway reactions/failure of cooling system or excess heating, etc. must be considered.
- This is very necessary when major modifications are being introduced.
- Standard operating and maintenance procedures
- Design and construction (mechanical integrity) of process units and machinery
- Provision of adequate internal refractory lining for protection of equipment shells.
- Pressure testing of process vessels, ducts, and piping.
- Structural beams and supports.
- Enforcing quality assurance/quality control (QA/QC) during repair of existing units and procurement of new units, if any.
- Human beings (check occupational hazards and health issues; ergonomic review of plant operations and control).
- Training and performance validation.
- Detailed instruction for safe working assigned to external parties and contractors.
- Safety devices, interlocks and audio-visual alarms.
- Electrical installations and interconnections for safety trips.
- Environment pollution control facilities for zero liquid discharge and minimizing air pollution.
- Environmental impact assessment on premises and surroundings likely to be affected.
- Preparedness for emergency response and its field verification.

9.5 Safe and Pollution-Free Start-Ups

9.5.1 Safe Start-Ups

Ideally a safe and smooth start-up of a plant after erection can take place when all precautions are taken and standard operating procedures are followed. It can result in the plant attaining the rated production capacity earlier with products as per specifications, generates minimum effluents, minimises wastage of raw materials and utilities and enables meeting supply schedules to customers. It also builds up employee morale and reputation of the organisation. Any planned stoppage (for maintenance) and restart of the plant can be done smoothly; and it can be run continuously as required in such cases.

However certain unsafe situations may occur due to sudden stoppage of a chemical plant (say, during power outage or breakdown of some key unit) and these should be kept in mind while restarting the plant. Additional safety precautions may have to be taken and the standard procedures may have to be modified while restarting the plant in such cases. Some typical situations are as follows:

- Sudden jerk can occur on high pressure units if NRV is not provided on connection to other high pressure units. This can cause safety issues or blow off from safety valves which can waste materials inside the process unit.
- Production of hot reaction products/hot gases can continue for some more time (even **after the plant has tripped**) since there could be considerable hold up of material inside the process units (like furnaces which are being operated as high temperature reactors), agitated reactors, pressure vessels, condensers, etc. The sudden stoppage of cooling medium to condensers or cooling jackets can pressurise the process units to dangerous levels. Blow off from safety valves can also occur.
- *Multiple Effect Evaporator:* A pressure build up may occur in the last effect when the circulation pump for the barometric condenser (final vapour absorber) trips. This can also disturb the flow of vapours from earlier effects.
- Absorption Tower: Example, there can be release of unabsorbed SO_3, NO_2, etc. when the circulation pump for absorption tower of a sulphuric acid/nitric acid plant stops, which can cause environmental pollution.
- Sudden stoppage of ID Fan: Certain chemical plants operate under negative pressure (below atmospheric pressure) which are provided with induced draft fans at the end to remove gases from the process units *by sucking them*; and then discharging through an exit chimney.
- There are possibilities of excess build-up of pressure or combustible vapours in combustion chambers or reactors; escape of uncondensed vapours from vents (if provided) or unabsorbed gases from the absorption tower in case of a sudden stoppage of the ID Fan due to any reason. This can create undesirable conditions like concentration of oil vapours reaching explosive limits, loss of production due to interruption in process, heavy environmental pollution, etc.
- Choking of pipelines: There are chances of choking of pipelines (particularly horizontal lines which are carrying slurries with suspended solids). Solidification of molten liquids may occur when the supply of heating medium gets interrupted. Hence provide drain valves at suitable locations for emptying out the pipeline in case a sudden plant stoppage occurs.
- Overflow from process tanks: An overflow from process tanks (which are at lower level) can occur when overhead tanks or process units at higher level get drained during power outage or sudden plant stoppage. Hence provide NRV at appropriate location in the pipelines to prevent backflow from overhead tanks or process units.

- Pneumatically operated drain valves for process vessels: These may remain open if supply of compressed air gets interrupted/reduced. This can drain out the entire process vessel. Hence provide valve keys with levers for manual operation/ attached operating chains for closing such valves.
- Jamming of Agitators: These can get jammed in melters, mixers, and dissolvers due to presence of un-dissolved/large lumps of solid materials.
- Other units: (1) Pusher centrifuge impeller can get stuck in slurry. (2) There will be chances of heavy weights or heavy equipment falling down unless automatic brakes can immediately prevent downward motion of electromagnetic hoists, EOT Cranes.

9.5.2   Preventive Measures

The plant engineers shall anticipate such matters and make advance provisions to deal with them. Some suggestions are given below as preventive measures/to enable easier restart of the units. These can be modified by the readers to suit their individual plants.

- Provide drain points at bends/change in direction of pipelines or provide a gradual slope for the pipeline so that they may get drained into the supply vessel.
- Design and construction of reactors with high safety factor.
- Provide emergency cooling medium supply for circulation in cooling jackets through alternative source (*see below*).
- Provision of properly designed and installed safety valves and rupture discs.
- Use thicker shells and better MOC for reactors and condensers.
- Controlled feed of reactants to avoid excessive hold up inside the furnaces, reactors, and condensers. Use metering pumps and rotameters.
- Provision of supply of cooling water/medium from dedicated overhead tanks (ESR—elevated storage reservoirs). These tanks should be always full with water and should have an outlet valve at the bottom which is to be programmed as " Normally closed, but to Open when power outage occurs in the plant".
- Make arrangement for automatic supply of power from DG diesel generator sets to key units or which can get jammed. These sets shall be programmed to start automatically when power outage occurs.
- Provision of ducts for release of gases/vapours through (alkali) seal pots through tall vent lines. The gas release vents shall be provided on key equipment.

Special Precautions
- These are to be taken during operation and maintenance and must be clearly understood by all operating and maintenance personnel.
- If possible, a "dry run" of the plant may be conducted.
- Assembly spot shall be clearly notified in the premises.
- Neighbouring units may be informed before the start-up if there is a chance of release of toxic gases or any excess effluents.

9.5.3 Pollution-Free Start-Up

- Check operation of effluent treatment and air pollution control facilities before starting the plant. Check solid waste disposal/safe collection arrangements.
- Confirm that the DG sets and firefighting units are in ready to use condition.
- The exhaust fans must have sufficient capacity to remove the air pollutants from the place of generation as well as working areas. The exit gases should be discharged away (at a height) through suitable scrubbers to prevent pollution in working areas as well surroundings.
- There could be escape of unreacted raw materials due to insufficient conversion (by the converter) till the process units attain ideal/steady state conditions. There could be escape of finished products in gaseous state if the condensers or absorption tower are yet to attain required conditions during the plant start-up. Hence the scrubbing units must be started first.

Hence first check and confirm that:

- Converter is at required temperature. If not, use auxiliary heating system to heat up the catalyst before starting the plant e.g. sulphuric acid plant start-up).
- Cooling system for condensers is working properly.
- Absorbing liquid is flowing properly at ideal concentration, pH, and temperature and is distributed in the absorption tower to irrigate the internal packings (gas liquid contact surfaces).
- Confirm that the gas inlet duct for absorption tower reaches up to the centre for proper distribution of gases in the tower.
- Confirm availability of standby pumps for absorption tower, venturi scrubber, packed tower, etc.
- Check operation of valve at throat of venturi scrubber and clean the pressure taps at gas inlet and exit. Adjust this valve during start-up. Valve in bypass gas duct should be closed fully. It is not to be opened during start-up of the plant.
- Check gas flow through sampling lines and sample conditioner.
- There shall be sufficient scrubbing liquid in the circulation tank and available stock in feeder alkali storage tank.
- Confirm working of scrubbing liquor pump(s). Flow through venturi scrubber and packed tower should be satisfactory.
- In case the gas stream contains particulate matter the flow of scrubbing liquor should be at least 1.0 L per actual cubic meter of gas flow **in the venturi scrubber**. It can be up to 3.0 L per actual cubic meter of gas flow. *However too much excess can only create high pressure drops.*
- In case the gas stream contains other pollutants the flow of scrubbing liquor should be at least 4.0 L per actual cubic meter of gas flow **in the packed tower** provided after the venturi. It can be up to 6.0–7.5 L per actual cubic meter of gas flow. *However too much excess can only create high pressure drops.*
- Confirm working of the system for alkali addition to scrubbing liquid for maintaining the pH/concentration of alkali at desired value.

- Bag filter: Check the compressed air back pressure systems and working of timer.
- Temperature of gases at inlet must be below limits given by manufacturer. Check pressure taps at the gas inlet and exit of the bag filter. Very low pressure drops indicate torn bags. Change the torn bags by new ones to prevent escape of fine dust.
- Electrostatic Precipitators: Check alignment of discharge and collection electrodes. Check electrical power circuits and control arrangements for voltage across the electrodes.

  Check the ash draining and removal system.
  Check working of air lock valve and gas pressures at the inlet and exit

- Operate the plant at as less a rate of production as possible (especially after any long stoppage) because the process units will not be at ideal conditions immediately after the stoppage. These are required to reach an ideal steady state whereby the possibility of pollution is minimal.
- The rate of production can be increased after confirming that facilities of air pollution are working properly and emission from chimney is getting reduced as the process conditions are getting normalised. *Example: Typically, this is to be done when a sulphuric acid plant is to be run slowly after a long stoppage.*

9.6 Managing Human Resources

- The plant management should carefully handle the human resources in the organisation in order to ensure safe, pollution-free, and efficient working of the plant.
- These resources could be classified as supervisory staff; operating and maintenance personnel; and auxiliary support personnel.
- Stores department, utilities section, firefighters, safety department, welfare department, security, marketing, finance, transport, etc. form auxiliary support services in the organisation.

9.6.1 Activities and Jobs

The following jobs are to be done by personnel in the organisation either themselves or with external assistance. All personnel in the organisation should work as a well-managed and coordinated team.

- Preparation of inquiry documents for procurement of process units and machinery (while making specifications to suit product mix and site conditions).
- Technical and commercial evaluation of offers from vendors.

- Preparation of fabrication drawings for equipment in-house if the organisation has technical expertise.
- Assistance to management for drafting purchase orders.
- Drafting terms for performance guarantee runs for individual units.
- Stage Inspection of equipment during procurement.
- Final Inspection of equipment prior to despatch from fabricator's workshop.
- Site fabrication of storage tanks, large-sized process units, and their testing.
- Finalizing and inspecting connecting ducts and piping.
- Erection of process units and assembly of internals at site.
- Matters concerning high pressure units.
- Final selection of material handling systems, rotary equipment, and machinery.
- HAZID and HAZOP studies and subsequent corrective actions.
- Trial runs of individual machines followed by sub-sections and dry run of entire plant.
- Commissioning of the plant.
- Compliance of statutory and safety instructions.
- Safe and efficient stabilised operation of the plant.
- Capacity expansion/diversification in future.
- Operation and maintenance of heat recovery equipment—waste heat recovery boilers, air pre-heaters.
- Refrigeration plants and cooling systems.
- Marketing of quality products as per schedules for dispatch. Attending to queries and complaints by clients.
- Planning and execution of maintenance of process units and accessories.
- Decisions for make or buy---spares, development of better equipment.
- In-house development of substitute units.
- Controlling inventory of spares.

9.6.2   Matters to Be Looked into While Hiring Persons

9.6.2.1   Qualifications and Experience

The persons should be suitable for handling the process units and machinery. They should understand the chemical reactions, heating/cooling required, and the physical as well as chemical properties of the materials to be handled. They should always be aware of the dangers involved and the safety precautions to be taken.

9.6.2.2   Physical Fitness for Working

The persons should be capable of going up and down at different floors to operate valves, charge in raw materials, to monitor process conditions by keeping a watch on the instrumentation and to operate essential machinery for process.

9.6.2.3   Alertness While Working in the Plant

Chemical industries need careful and continuous watch on the process units and machinery running in the plant. Material feeding rates for reactors, flow rates of process streams, and operating levels in various tanks and reactors; pressures, temperatures, pH, load on various drives need constant attention. It needs persons who are mentally and physically alert always.

9.6.2.4   Working Conditions for Efficient Operation

Management shall always try to make the working safe and convenient for the personnel.
Key equipment must be provided with:

- Adequate lighting (which should be flame proof in areas where inflammable vapours or gases are present) to observe instrument dials, control valves, and level indicators of tanks (which may overflow if not carefully attended).
- Provision of ID fans/ventilation fans in working areas to remove dust, chemical vapours, and gases escaping from process unit during charging of reactants.
- Easy-to-use stairs, railings, and work platforms away from furnaces.
- Immediate access to safety devices and safety showers.

9.6.2.5   Remuneration and Other Facilities

These shall be commensurate with job content, responsibility involved, experience gained, and sincere efforts to meet organisation goals for safe, pollution-free operation to manufacture quality products at required production rates.

9.6.2.6   Safety and Training, Refresher Courses

Regular training sessions shall be conducted for the operating and maintenance personnel. These shall incorporate revised instructions from OEM, and upgrade them after a review of HAZOP. This should be preferably done every 4–6 months.

9.6.2.7   Incentives Can Be Given for Efficient Working

Employees can be given incentives for productive suggestions, for increasing safety of personnel and plant assets, for reducing cost of production, for improving product quality, for better utilisation of assets; for reducing downtime, inventory of spares and maintenance expenses. There should be no favouritism in award of incentives.

9.6.2.8 Quality Circles

These should be encouraged for improvement in the plant units and systems. Regular follow-up on suggestions should be carried out. Good suggestions and efficient actions shall be rewarded in a transparent manner.

9.6.2.9 Disciplinary Action

Rules and regulations should be well defined and informed to all concerned. Any disciplinary action being initiated shall be with a warning for a misconduct, with an opportunity to explain the misconduct. The action to be taken shall be proportionate to the likely consequences of the misconduct committed. *Avoid excessive punishment.*

9.6.3   Provision of Protective Equipment

Provision of such equipment and facilities to all working personnel and those who are likely to be present in the concerned area can build up employee morale and improve monitoring of process conditions as well as attention to any problem in the plant.

- Face shield and safety helmets
- Acid- and alkali-resistant dress
- High temperature-resistant dress
- Full length hand gloves and gumboots
- Safety shoes with internal metallic protection
- Special shoes with rubber soles (to prevent generation of sparks)
- Portable breathing apparatus and oxygen cylinders
- Gas detectors at strategic locations
- Gas masks with appropriate canisters
- Rope ladder, whistle, safety lamps for working inside closed vessel
- Safety showers and eyewash fountains with water supply from dedicated overhead tank
- First aid facilities with a dedicated medical team 24 × 7 should be available in the premises. It should be easily accessible from all working areas in the plant. This facility should be well stocked with all necessary antidotes against all chemicals being handled in the plant.

9.6.4   Span of Control

It is essential to have a very safe and spacious plant layout for proper movement of materials and personnel; to keep watch on key equipment; to monitor operating conditions and to quickly take corrective actions. Operating personnel can keep watch on more equipment in a better manner with a *typically well designed* layout of process units and machinery at all levels of the plant. The layout shall have:

- Convenient location of control room.
- Ease of observation of key equipment, instrument dials, and peep holes; easy convenient approach to all operational valves and process control systems by the operating and supervisory personnel.
- Main fuel tanks and storage tanks of dangerous chemicals must be sufficiently away from working area. Statutory requirements must be complied with.
- Free passages and escape routes between process units must be maintained. There shall not be any hindrance due to any reason.
- Space available around machinery for ease of proper maintenance.
- Convenient location and orientation of cleaning manholes for quickly draining out residual matter and for cleaning the unit.
- Quickly accessible diesel generator (with sufficient fuel capacity to run key equipment for about 8 h at least) for emergency power supply.
- Sufficient number of safety showers and gas detectors should be installed in areas handling dangerous, toxic, and inflammable materials
- Control valves should be located away from hot surfaces and sources of gas leaks.
- Ladders and work platforms for ease of cleaning and maintenance.
- Provide protection from rains and adverse weather conditions.

9.7 Engagement of Contractual Services

Unskilled/semiskilled contractual labour is generally employed for repetitive work during running of the plant; for removing tower packings, during major repairs, during annual overhaul, etc. The engagement needs to be carefully considered and all applicable laws of the land should be complied with.

- Jobs which need large manpower; for which very specialised know-how is required; for which special tools and other resources are required or are having large voluminous work may be awarded to outside parties (contractors having skilled manpower or sufficient resources).
- Detailed description of jobs to be done must be explained to the contractor. Battery limits of scope of work including initial conditions of units (before assignment of the work) and the final condition of the unit expected after the work is finished must be clearly defined.

9.7.1 Examples

– Completely taking out the tower packings, cleaning, and filling back with make-up quantities to replace the damaged packings.

- Complete removal of sludge from a storage tank and thorough cleaning from inside.
- Expected performance from a big pump or fan after repairs.
- Repairs to refractory lining after removal of damaged portion.

9.7.2   Special Requirements

Jobs to be done in closed vessels, at a considerable height, and in the presence of dangerous materials which involve risk shall be properly defined. Safety officer must be consulted before award of contract. Decontamination of process units and adequate safety arrangements must be done including provision of first aid facilities.

Contractors must have accident insurance policies for all workforce employed by them.

Facilities generally required by contractor shall be provided immediately to prevent contractual labour lying idle.

Rules and restrictions for movement by contract workers in the premises must be made clear.

Any job not being done properly/abandoned without proper reason or not done as per instructions from plant engineers may be carried out through another contractor and the expenses shall be recovered from the payment due to the original contractor.

Typically, controversy arises when the contractor uses substandard materials of construction MOC for protective lining or material which is not tested and approved by the plant engineers.

In case of only job work, controversy can arise if the material (given by the client) is to be brought from the storage located at a considerable distance from the job site. This needs extra effort by the contractual labour. Hence scope of work/battery limits shall be clearly defined.

Site should be thoroughly cleaned after the job is completed.

All necessary tools and equipment must be brought by the contractor and should not depend on the client for them.

9.7.3   Differences Arising During Contractual Work

Such matters should be settled amicably between plant engineers and contractors.

- Process unit/machinery handed over for maintenance is either not properly isolated or residual dangerous materials are not fully removed by client.
- Storage space, security, and other site facilities not provided to the contractor.
- Volume of work is more than what is agreed as per the contract.
- Volume of work/difficulty has increased substantially while the job is in progress.

This can happen when more repairs are required than expected.

- Completed job not tested/trial not taken due to idle plant or raw materials not available or there is little demand for the product; *but contractor is asking for payment.*
- Safety arrangements are not adequate.
- Power, water, steam, fuel, and more unskilled labour is not made available to the contractor even after requests.
- Contractor has interfered with other process units or movement of materials in the premises.
- Some process units or machinery has got damaged due to careless work.
- Contractor is insisting on stopping some other units or machinery on safety issues or some other reasons which can disturb operations in the plant.
- Contractor has diverted his manpower elsewhere and thus delaying the job given on contract. This is likely to delay restart of the plant.

9.8   Guidelines for the Project Engineers, Plant Engineers, and Maintenance Engineers

9.8.1   Guidelines for Project Manager

Following matters should be clearly understood and executed for the successful implementation of a project.

- Objectives for the project shall be clearly defined i.e. whether it is for expansion of capacity/modernisation of existing plant or diversification to more products or the revival of an idle plant.
- Proper planning for implementation of scheduled activities shall be done.
- List of long delivery items and the dates planned for delivery to site shall be known.
- Final selection of Vendors/Fabricators shall be done before commencement of job.
- Dates informed by finally selected vendor for stage inspection, final inspection, and delivery to site shall be as per implementation schedule. Deviations from purchase orders proposed by vendor and agreed by the purchaser must be known.
- Copies of orders placed on equipment vendors, site fabricators, and local contractors shall be given to project manager.
- Regular follow-up with these vendors and suppliers for delivery of the items shall be carried out.
- Obligations of the buyer and vendors, site contractors are to be defined clearly; as well as battery limits of scope of supply.
- Arrangements to be made at site for development of land and site infrastructure (office, security, stores, power for construction, workers' amenities, lighting, first aid, transport, and water required for construction).

- Construction of civil structures and foundations for various process units, machinery, and storage facilities shall be done as per layout finalised for the project.
- Project manager and erection engineer should have authority for engaging additional manpower for site work if necessary.
- Copies of Assembly drawings of Process units and nozzle orientations for all process streams; P &ID for the plant should be given to project manager.
- Conducting HAZID and HAZOP studies by designer and experienced operation and maintenance personnel. Project manager should be given copies of the final recommendations in order to make required changes in the piping, ducting, etc.
- Details of electrical installation.
- Conditions to be fulfilled for obtaining all statutory permissions (for fuel storages, pressure vessels, boilers, economisers, super-heaters and steam lines, provision of safety devices, escape routes from dangerous areas).
- Compliance with any additional conditions and instructions from statutory authorities to be carried out for consent to establish the plant.
- Copy of performance guarantee for individual equipment and for the entire plant should be given to project manager.
- Instructions manual from OEM for safe storage, erection, lubrication, mechanical trials, and commissioning of key equipment purchased.
- Keep a reasonable fund always available with project manager for emergency purchases of items at site so that the project is not delayed.
- Keep track of daily progress of work, expenses, and the cumulative expenses incurred on project; and possibility of exceeding the sanctioned amount.

9.8.2 Examples

Guidelines for Erection and Commissioning of (Gas–Liquid Contact) Tower Typically Used as an Absorber Tower
- Bottom plate of tower must be provided with corrosion protective paint from outside.
- A removable duct length should be provided in the gas inlet piping near the bottom of absorption tower. Gas entry duct shall be downwards in to the tower. An extension of gas inlet duct up to the centre of tower to be provided for proper distribution of gases.
- Check levelling of tower internals like liquid distributors, supporting raschig rings at bottom and the next few layers, demister pads, and fiber bed mist eliminators.
- Only a few layers (three to four) of the rings shall be arranged at the bottom, rest shall be randomly packed. They should not be dropped from a height, but shall be gently released from a container of suitable size.

- Liquid exit nozzle should be of sufficient diameter, and limbs of the U-bend in the exit piping shall be of unequal heights for smooth drainage from the tower. This is to minimise hold up of liquid at bottom of the tower.
- Flow of absorber liquid shall be checked at the distributor provided at inlet as well as from the exit piping at the bottom. The level of liquid in the circulation tank may go down a little as the circulation is started on trial. This represents the liquid which has wetted the tower internals.
- After this, the level in circulation tank shall remain constant indicating no abnormal hold up in the tower.
- The current drawn by the circulation pump motor shall be constant; and there shall be no abnormal noise or vibrations.

Guidelines for Commissioning of Boilers

- Confirm that all necessary permissions have been obtained from statutory boiler inspection authority for the boiler, fittings; steam and feed water piping, etc.
- Run the feed water pumps on trial and confirm that they will be able to supply sufficient water to the boiler.
- For taking trial run of the feed pump: Throttle valves in the feed water lines to create back pressure (corresponding to the maximum operating pressure of the boiler) on the pump. Measure the rate of flow and the current drawn by the drive motor. The motor shall not get overloaded/trip. This trial run is to be taken for the standby boiler feed water pump also.
- Generally, the boiler tubes are given a protective coating of corrosion-resistant oils by the manufacturer. This should be removed before taking the boiler into service for steam generation as the oily layer can reduce heat transfer efficiency. This is done by carrying out an operation known as alkali boiling out. Dilute alkali solution is filled up in the boiler and slow firing is done. Vent valve on boiler is initially kept open to blow off some steam. It is then closed after a certain time and boiler pressure is raised a little. Firing is discontinued after this. Water inside the boiler is blown down and analysed for oil content. The operation is repeated till the oil content in blown down water is generally below 1 ppm or the limit prescribed by boiler manufacturer. However, instructions from boiler manufacturer shall be followed for the actual procedure to be followed.

9.8.3   Other Guidelines

- Confirm correctness of load data informed to the structural engineer.
- Provide additional reinforcement to support structures if the present ones appear inadequate (in case rotating/vibrating machinery is to be installed on the structure).
- Connecting pipelines and ducts should be separately supported to avoid their weights getting transferred to process units/pumps/machinery.
- Equipment is not to be installed at much greater height than planned.

- Thermal insulation and cladding to be provided as per internal temperatures, expected surface temperatures, and site conditions. *Too thin insulation can result in more heat loss. Any ingress of moisture through imperfect cladding can cause heat loss and corrosion of vessel/ducts.*
- Take proper earth connections during welding jobs.
- Remove temporary supports after the job is over.
- Clean the site of all debris after the job is over.

9.8.4 Guidelines for Plant Manager

The plant managers are experienced chemical engineers responsible for the day-to-day activities. They are required to run the plant in a safe, pollution-free manner and manufacture the products as per specifications to meet delivery schedules committed to the customers. These senior persons should always anticipate/keep in mind what can go wrong during erection, commissioning, or operation of the plant. They should take preventive actions and keep arrangements ready to control the situations like gas leak, fire, and spillage of corrosive, toxic, inflammable materials, if they occur.

They should also carefully look in to the recommendations of the technical team appointed for Hazop study and should plan for the prevention, detection, control, and mitigation of any hazardous situation.

They shall check the following regularly (every day):

- Availability of personal protective equipment and protective dress to all.
- Confirm satisfactory working of safety devices and interconnections.
- Confirm availability of all key personnel for process, maintenance, and instrumentation, (or trained substitutes in their absence).
- Stock of finished products of marketable grade.
- Stock of raw materials, fuel, and water (raw water, treated water, boiler feed water, etc.).
- Availability of all standby equipment, pumps, blowers, power supply arrangements (DG Sets) and sufficient stock of diesel to last for 16—24 h run of the DG set.
- Availability of necessary test apparatus and chemicals in process control laboratory.
- Condition of key process units and machinery; and whether they can be run at rated capacity/ or whether any of them should be run at lower than rated capacity due to some problem (high pressure drop, operating temperatures exceeding safe limits, excessive generation of pollutants).
- Working of gas detectors in sensitive areas storing raw material, finished products, and fuel storages.
- Condition of ETP, air pollution control units, firefighting arrangements, first aid arrangements, availability of emergency transport.

- Delivery orders/schedules for finished products given by marketing department.
- Storage, packaging, weighing, and transportation/dispatch system for the products.
- Production planning as per instructions received from higher management.
- To issue revised/modified instructions for operating personnel accordingly.

9.8.5 Guidelines for Plant Engineers on the Shop Floor

The engineers on the shop floor shall have a ready reference list of the following equipment and monitor their condition regularly. These should be in good working condition always. They should also regularly check and attend other items as given.

- Those operating at high pressures.
- Those operating at high temperatures.
- Those handling dangerous, toxic, or inflammable materials.
- Those equipment which are *not having any standby unit.*
- Effluent treatment plant and air pollution control units.
- Those units consuming large power.
- Maximum sanctioned power demand.
- Rotary equipment in the plant and speed control of agitators.
- Important Static units (storages, filters, etc.).
- Check deviations from standard/prescribed operating conditions for all process units and process streams (especially deviations which can create dangerous situations; deteriorate product quality; generate excessive pollutants or affect rate of production). Take corrective actions in case of deviations.
- Monitor quantities of raw materials issued and quantities of products obtained every day (if possible.) This should be done at regular intervals, say every week at least.
- Material balance shall be worked out accordingly to check the plant efficiency (conversion of raw materials to finished products; loss of raw materials; effluents generated).
- Monitor utilities consumption during plant start-up and during steady run of the plant. The above will be applicable to individual process units like reactors, evaporators, distillation columns, filtration systems, dryers, etc. to name a few. The analysis and feed rates of all inputs (materials and utilities) to important process units shall be regularly monitored. Corrective actions shall be taken at the earliest.
- Monitor equipment which need statutory permissions to install and to operate; and the last working day permitted (pressure vessels) by statutory authorities.
- Provide refresher training to shop floor work force. Check whether they are alert and attentive always by taking surprise rounds of the plant.
- *For Reducing breakdowns: While running the plant the operating technicians and shop floor engineers should be vigilant and should immediately try to find*

out the source of any abnormal noise, unusual smell (indicates some leak of process fluids), vibrations of pipelines and connected units (indicates problems with machines), development of hot spots on shells of furnaces or process units (as it can indicate damage of internal protective lining). Timely corrective actions in such cases can prevent major breakdowns.

- Process monitoring and control instrumentation. Confirm the accuracy of instruments. These shall be correctly calibrated and tested.
- Response time of process controller.
- Laboratory facilities and apparatus for testing the raw materials and finished products; and analysing process streams to monitor the process at important stages.
- Check the chemical analysis and physical properties, lump size, moisture content, presence of tramp materials, etc. of all raw materials procured.
- Ensure that the facilities necessary for proper product quality are in working order. They will include equipment for conditioning of raw materials and their purification such as screening, magnetic separation of tramp materials, drying, filtration, crushing pulverising, etc.
- Accuracy of raw materials feeding system.
- Facilities for purification and conditioning of products. They can include filtration, distillation, drying, addition of stabilisers, anti-caking agents, etc.
- Safe storage of raw materials in a well-covered place to protect them from rains, dust, vapours from processing units, etc. Non-compatible items shall not be stored together.
- The handling facilities (belt conveyors, hand trolleys, charging vessels) should be able to issue the materials from the storage to the processing section as per required rates. It should be possible to measure the quantities issued accurately in order to determine the consumption.
- Any unsafe condition in the plant; spillage or leak of process stream which needs to be attended immediately.
- Examine the protective lining of reactors, agitators, evaporators, heat exchanger, gas ducts, MOC of piping, and process units at regular intervals since the products can get contaminated if the linings are damaged.
- Working of water treatment plant and stock of treated water (soft water, demineralised water). *Ensure recovery of steam condensate.*
- Firefighting facilities.
- Condition of important process units (reactors, absorbers, condensers, etc.).
- Condition of important equipment for utilities including standby power supply units.
- Those needing statutory permissions to install and to operate; and the last working day permitted (pressure vessels) by statutory authorities.
- Provide refresher training to shop floor work force. Check whether they are alert and attentive always by taking surprise rounds of the plant.

9.8.5.1 Some Typical Examples of Process Systems

These are given here as practical guidelines for monitoring by the shop floor engineers:

At Inlet and Exit of Reactors
- Composition, temperature, pressure, pH (if applicable), and flow rates of incoming and outgoing process streams;
- Pressure and temperature of the reaction mass; Flow of cooling/heating medium.
- Operating levels, any leak from shell of reactor, abnormal noise or vibration of agitator, and current drawn by drive motor.
- Estimate conversion efficiency of catalytic reactor by comparing the working with ideal performance. Take out samples of catalyst to check if any deactivation has taken place.
- Whether the unit is operating within safe limits.

Filtration System
- Composition, temperature, pressure, and flow rate of incoming and outgoing process streams.
- Waste/impurities separated.
- Any loss of products with the impurities (can occur due to damage to filter media).
- Any sudden reduction of pressure drop indicates channeling/damage to filter media.

Evaporation System
- Concentration of dissolved solids in the process streams at inlet and exit.
- Flow rates of incoming dilute solution and outgoing concentrated solution.
- Consumption of heating medium (e.g. steam, furnace oil)
- Working of steam traps
- Condition of thermal insulation (monitor loss of heat by thermal imaging of shell).
- Operating levels of liquid in the unit and any spillage.

Rotary Dryer
- Analysis and flow rates of incoming and outgoing process streams.
- Whether outgoing product are getting dried as required.
- Operating temperature inside.
- Consumption of heating medium.
- Speed of the rotary dryer.
- Temperature of the shell.

Fractionating Distillation Column
- Analysis and flow rate of incoming liquid.
- Analysis and flow rate of each outgoing stream.
- Temperature and pressure of vapours at inlet of reflux condensers.
- Operation of reflux controls.
- Level of liquid in the fractionating column.

- Temperature and pressure of heating medium (generally steam).
- Working of safety devices.
- Working of steam traps.
- Consumption of heating medium.

9.8.6   Standard Operating Procedures for Process Workers

Operating teams generally consist of production engineers, shift supervisor, senior operators, helpers, and contract labour (if hired occasionally). They should be fully familiar with the following. *Regular refresher courses shall be conducted if they are not familiar.*

- Instructions for safely starting and stopping the plant.
- Dos and don'ts during operation to ensure safety and efficiency.
- Important process parameters to be observed at regular intervals.
- Corrective actions to be taken in case of deviations from set values.
- Emergency procedures during power outage or breakdown of key equipment.
- Operating instructions for increasing/decreasing rate of production.
- Safe methods for handling dangerous materials.
- Safety precautions while attending to fires; major leaks of process fluids.

9.8.7   Guidelines for Maintenance Engineers

It is a must to carry out maintenance jobs immediately if there are unsafe conditions or if statutory instructions have been received.

Maintenance teams generally consist of maintenance engineers (mechanical and electrical), fitters, welders, riggers, electricians, instrument technician, and contractual teams (outside parties).

Maintain all tools for mechanical, electrical, instrumentation and civil repairs, cutting and welding sets, chain pulley blocks, slings, etc. in good working conditions.

9.8.7.1   Standard Safety Procedures for Maintenance Workers

These include the following:

- Complete isolation of unit (which is to be attended) from connected units.
- Complete removal of residual material and thorough cleaning of the unit before taking up maintenance jobs by taking help from process workers.
- Collection/neutralisation and safe disposal of waste generated.
- Check presence of residual material by suitable detectors.
- Arrange firefighting equipment and keep the firefighting system activated. Check stock of water, CO_2, and dry chemical powder cylinders.

- Arrange all necessary tools for lifting and repairs, cutting, and welding.
- Provide separate earth connections for welding.
- Use ventilation fans and personal protective equipment while carrying out maintenance jobs.
- Follow the instructions from original equipment manufacturer.

9.8.7.2 Important Activities by Maintenance Engineers

- Regular checking/testing and confirmation of working of all safety devices, safety showers, firefighting system, and electrical safety interconnections (as well as audio- visual alarms and trip systems) along with process operating personnel.
- To ensure all standby equipment, pumps, blowers, filter, material handling systems, etc. are in ready to use state.
- Regular checking of rotary equipment for lubrication, vibration, and abnormal noise.

Example *Check speed, suction and discharge pressures; and the current drawn by drive motor (indicates electrical load) of ID Fan regularly in addition to the monitoring being done by process personnel.*

- *Drain out accumulated condensate from gas ducts and body of ID Fan, etc. every hour.*
- *Check continuity of earth connection/lightning arrester provided on the chimney.*
- *It should be possible to climb up to check the chimney, guy ropes, ground foundation bolts, sampling lines, etc.*
- *Physical condition of self-supporting chimney shall be checked once in 3 months.*
- Coordination with stores, marketing and production departments for preventive maintenance programme.
- Regular follow-up with procurement section for maintaining sufficient stock of key spares.
- Development of in-house schemes for increasing life of process units and machinery; as well as costly spares.
- Planning for annual overhaul activities and major repairs.
- Guiding operating personnel for safe and smooth running of plant machinery.
- Assisting senior management to draft purchase orders for new units.

9.8.7.3 Situations When the Annual Shutdown May Be Postponed

- When the plant is running smoothly in all respects (rate of production, product quality, consumption of raw materials and utilities within limits, control on effluents, etc.) and *permission from statutory authority is available* for continuing operations of plant units and machinery (operating at high pressure/higher loads).
- When there is steady demand for the product.
- If an important/urgent order for the product is pending.

- If procurement of spare parts is incomplete.
- If it is planned to carry out major repairs or replacement of unit; and preparation (arrangements for special tools, new unit from fabricator, etc.) is yet to be done.
- When it is desired to observe the effect of modified operating conditions on the performance of the plant.
- If it is desired to compare product quality and performance of the plant when raw materials from different sources are to be used.

However postponing maintenance unreasonably can result in:

- Loss of efficiency of plant.
- Waste of raw material, energy, and water.
- Poor product quality.
- Generation of excessive effluents.
- Development of unsafe situation.
- Corrosion and reduced life of equipment and adjacent units due to leaks of corrosive chemicals.
- However, the plant should not be run beyond the last permitted date for operation of pressure vessels.

9.8.7.4   Guidelines for Advancing the Shutdown Before the Due Date

- If it has become unsafe to operate the plant unless urgent repairs are carried out.
- If there is excessive generation of effluents.
- When consumption of raw materials and/or utilities have increased too much beyond reasonable limits.
- When product quality has deteriorated.
- When rate of production has reduced considerably and cannot be improved unless major repairs are carried out.
- If there is very little demand for the product.
- When cost of production has increased beyond acceptable limits.
- When raw materials are not available currently as per specifications; but will be available after some time.
- If a key process unit/machine is not performing well and it needs to be urgently attended by stopping the plant.
- Key tools and other facilities required for major repairs will not be available later on due to some reason.

9.8.7.5   Safety Steps for Maintenance Work on a Process Unit or Plant Equipment

Obtain work permit from senior responsible engineers and safety officer that following steps have been taken:

- Complete isolation from other connected units. Insert blinds in piping as a matter of abundant safety (in case of a chance of valves in the connected piping passing through i.e. leaking through).
- Decontaminate thoroughly by draining out safely all residual material inside and flush by water/dilute alkali as applicable. Collect the waste liquors and send to ETP.
- Vent out/suck out any residual vapours inside by exhaust blowers. Discharge through scrubbers/seal pots with dilute alkali and release at a height away from working area. Check and confirm by gas detectors that all vapours from the unit are exhausted.
- Keep ready firefighting system, portable breathing apparatus, and air circulation fans.
- Provide separate earth connections during welding.
- After the maintenance jobs are over, test the process unit/machinery as per standard procedure before taking into use. Any leak/vibration observed must be attended.
- Confirm that all maintenance personnel have finished the jobs allotted and are away from moving machinery/have come out from process units.
- Issue work completion report (clearance to start the units) after all jobs are over.

9.9 Suggestions for Reducing Breakdowns

Working of a process plant is disturbed when some process unit or plant machinery breaks down. Effect of such disturbance is felt more in case of a continuously operating plant. It can affect product quality, can generate more effluents or create unsafe conditions. It is therefore necessary to prevent breakdowns in the plant.

Some typical actions to prevent breakdowns are:

9.9.1 At Design Stage

- Duty conditions are to be specified correctly while designing the equipment/ while fabrication drawings are being finalised and released for fabrication.
- Sufficient margins must be provided to take care of occasional deviations from specified operating conditions (pressure, temperature, pH, flow rates). Correct information about the corrosive, toxic, and explosive/inflammable nature of materials to be handled should be given to the equipment designer.
- Some additional margins may be added to prevent breakdowns due to failure of controlling instruments, operator error, internal leaks of process streams, etc.
- Provide water cooled glands or mechanical seals for rotating shaft seals instead of simple gland packing which may leak.

- Drive systems with directly coupled drives or drive through belts/gear boxes as per load conditions should have accurate alignment of rotating parts.
- Strengthening the equipment/machinery parts in contact with corrosive, erosive materials; those required to operate at very high speeds, at high temperatures, fluctuating process conditions, in outdoor installations, or subject to dusty atmosphere, or to corrosive vapours. The materials of construction are to be carefully selected by first testing a small piece in working conditions.
- Provide intermediate bearings or supports for long rotating shafts.
- Provision of strainers in piping for process stream to remove erosive particles.
- Make provision of standby cooling arrangements.
- Provide good anti-vibration base pads.
- Provide additional reinforcements to structural supports and all weak components for improving equipment life.
- Design independent safety arrangements (safety vents and valves, rupture discs, safety trip interconnections to stop feeding of materials) and audio-visual alarm systems.
- Provide safety valve and rupture disc on incoming steam line connected for heating jackets of process vessels even when a pressure reducing station is installed.
- Oleum, SO_3 lines need to be kept warm. Hence, they have to be provided with steam tracing instead of jacketing. *A steam leak going inside the oleum/SO_3 streams can cause runaway reaction and generate excessive pressure.*
- The inlet lines of heating media shall not be of too large size (for proper control of flow of heating media). An impingement plate must be provided in front of the steam inlet side of the jacket to prevent direct impact of high velocity steam on the vessel wall which is being heated.
- Cooling media should be circulated through large diameter lines to control temperature inside the reactor/process vessel. Alarm shall be provided if flow of cooling media stops.
- On-line analysers for important process parameters and all effluent streams.

9.9.2   During Procurement

- The operating conditions shall be specified as close to actual values as possible (*with some upper and lower deviations within practical limits*) while obtaining offers for equipment. This will enable the equipment to be designed and fabricated more accurately in order to get an efficient performance. The fabrication of equipment is to be carried out after considering both upper and lower limits of operating conditions. *However, specifying of very wide tolerance limits of dimensions, pressure and temperature ratings, etc. is to be avoided.*
- Quality assurance plan (QAP) conditions must be defined carefully in details. Include clauses for certification/testing of MOC and stage and final inspection by purchaser's technical representative.

- Specify correct welding process, use of compatible electrodes, ASME Sec XIII and IX codes for fabrication, power settings (AC/DC), requirement of radiography, post weld heat treatment (PWHT), and hydraulic tests for pressure vessels and also for those process units required to handle dangerous, lethal materials. *Do not waive any conditions as it can make the equipment unsafe to operate.*
- Insist that all required radiography tests, PWHT, and non-destructive tests are to be done as per statutory requirements; and are to be witnessed by technical representative of purchaser and inspecting authority. Preserve records of all such tests.
- Stage inspection or final inspection should be done as per statutory requirements. No conditions shall be permitted to be waived by unauthorised persons.
- Final test done to be done with chloride free treated water. Do not permit use of raw water instead of treated water. No traces of water should remain inside the vessel before trial runs or commissioning of the plant.
- Instruments used for the tests such as pressure gauges, dial thermometers, thickness gauge must be pre-calibrated and certified for accuracy.
- Radiation protection must be available to all concerned during radiography tests.
- Do not permit any changes on the equipment by cutting and/or welding after PWHT.
- MOC used for fabrication of wetted parts must be tested or certified for composition, corrosion resistance, mechanical strength, and other necessary properties.
- Non-standard (bought out) components should not be used during assembly.
- Test certificates should be available for bought out components like safety valves, gauge glasses, check valves, blow down valves, etc.
- Nozzles on equipment should not be used for tying by chains and then lifting the equipment. Use only lifting lugs while lifting and loading on transport vehicle.
- Keep the equipment on wooden supports during transportation.
- Close all openings by blinds during transportation to prevent ingress of dust, moisture.
- Provide proper weatherproof protective covers during transportation and storing at plant site.

9.9.3 During Erection

- Dimensions of foundations and positions of foundation bolts should match with base plate of equipment/drawings given by vendor.
- Civil Foundations must be sufficiently cured before placing equipment on them. Clearance must be obtained from civil engineer.
- Corrosion-resistant treatment/paint should be provided on base plates and structural support columns and beams
- Thermal expansion joints should be constructed at appropriate locations internally during refractory lining work

- Provide expansion joints/bellows in ducts (which will be subjected to considerable temperature changes).
- Carefully check all joints and possibilities of leaks from vessels, shaft seals, glands of valves, etc. Attend immediately prior to mechanical trials and commissioning of plant.
- Provide strainers in sampling lines of pH meters, conductivity probes, and suction lines of metering pumps.
- Cast iron lines and glass lines must have adequate supports. These shall not be subject to tensile forces during (and after) erection.
- Sufficient clear space must be available between equipment and building walls, other adjacent units for proper fitting of valves, cleaning manholes, instrument probes, and for cleaning during any shutdown.
- At least two or three independent escape routes and sufficient clear space must be available near process units and machinery handling dangerous materials.
- Provide proper work platforms, ladders, and railings for process units.
- Provide proper (flame proof if required) lighting near valves, instrument dials, etc.
- Roller (or sliding) supports should be provided for linear expansion of long units (which will run at high temperatures when the plant is operated) such as refractory lined horizontal furnaces.
- Each layer of high temperature protective/acid-resistant lining must be given sufficient setting time and should be cured properly before commissioning.
- Vents are to be provided for release of moisture/vapours generated during initial heating of the lining or process units. The gases should be released at a safe distance away from working areas.
- Follow procedure recommended by refractory manufacturer for refractory drying and curing. Use only clean treated water for preparing cement paste.
- Drain valves and lines to be provided to drain out water inside the process vessel and equipment before trial runs.
- Make provision for flushing out ducts by suitably connected water lines and drain lines leading to effluent treatment.
- Check correctness of orientation of nozzles for connection to incoming and outgoing process streams which must match with P& ID and process flow.
- Confirm proper alignment of driver (motor, gear box) and driven units (pump, compressor, blower, agitator shafts), sufficient length, curing and setting of foundation bolts; provision of lock nuts or cotter pins (which are necessary to prevent loosening of foundation bolts *due to vibrations*).
- Charge in the correct grade and quantity of lubricants as per recommendation by OEM. *It should be changed after a few hours of trial runs (as recommended by OEM) or as per analysis of the samples drawn.*
- Provide adequate arrangements for cooling of lubricants before trial runs.
- Lube oil pumps to be provided with emergency power (battery backup).
- Provide drive motors of correct power and speed rating as recommended by OEM
- Examine all motors, electrical installations, control panels, electrical instrumentation, interconnections, relay settings for safety trips, etc. before mechanical trials.

- Provide power supply as required by the motors (voltage, frequency).
- Confirm earth connections of vessels and piping handling inflammable, explosive materials for continuity, and current carrying capacity.

9.9.4   During Testing of Process Units and Mechanical Trials of Machinery

- The units and machinery should be cleaned by air flow (or washed if permitted) after erection is completed to remove scales and dust before testing or taking mechanical trials.
- Instructions from OEM must be followed during testing the process units and plant machinery and while carrying out hydraulic pressure tests.
- Smooth operation of all valves is to be checked along with external indicators of the spindle/flap position (which must correspond correctly with the internal position of the flap). Locking arrangement should be available for the operating lever.
- Confirm that correct grade and quantity of the recommended lubricant has been charged.
- Free rotation of rotors shall be first checked by hand or by wrench instead of starting the motor directly during mechanical trials.
- The trial runs should not be stopped abruptly. Any problem must be addressed immediately before starting the production runs of the plant.

9.9.5   Before and During Commissioning Runs

- Confirm settings of all safety devices, interconnections, alarms for high/low levels, temperatures, pressures, etc.
- Confirm availability of safety showers, firefighting units, and personal protection equipment; conveyance, first aid facilities, etc.
- Check calibration of all instruments.
- Confirm availability of raw and treated water, fuels, diesel generator sets, sufficient quantities of raw materials and other inputs, laboratory test facilities, etc.; and presence of experienced, trained operators, and maintenance crew; maintenance facilities and adequate spares for machines and electrical systems.
- Preheat the process units (if required for start-up) as per standard procedures.
- Start the process units on low loads as per approved procedures.
- Check for any leaks, abnormal noise, vibrations, heavy current drawn by motors, temperatures, pressures, flow, concentrations, etc. of all process streams and process units, generation of effluents/exit gases from chimney.
- Feed in measured quantities of raw materials and check all process parameters for different process units as well as output quantities.

- Slowly increase production rate when the process units achieve safe stabilised run and while the product quality is acceptable; while environmental pollution is controlled.
- Always operate below maximum rated capacity/below maximum permissible conditions of temperature, speed, flow rates, pressure, etc.
- Reduce drastic operating conditions of pressure, temperature, concentration, and flow rates in process vessels if possible (without affecting output or product quality).
- Provide more standby units and consider operating the installed standby units regularly (say, in alternate week).
- Simplify design and construction by giving practical suggestions to designers and fabricators. These can include suggestions for strengthening the wetted components/weak parts by more robust construction; use of more corrosion-resistant MOC
- Also consider standardisation of design of spares for similar machinery to reduce inventory.
- Provide strainers/filters for removing the erosive particles from process streams to protect the impellers of pumps, agitator shafts, and lining material of process vessels.

9.9.6   Check and Confirm

- Proper location of pressure, temperature taps, and sampling nozzles in ducts/pipelines.
- Grade and material of filter cloth, mesh size of filtration screens.
- MOC and thickness of gaskets in all bolted joints of process vessels and piping carrying dangerous, toxic, and corrosive inflammable fluids.
- Thickness of flanges for vessels and piping carrying dangerous lethal materials or at high temperatures and pressure should be as per standards.
- Recycle line must be provided of sufficient diameter for positive displacement type pumps, BFW pumps. These should be as per recommendation of OEM.
- Provision of anti-vibration pads must be as per specifications for machinery. This is to prevent disturbance of settings of control valves, loosening of foundation bolts, vibration of glass/acid-resistant lining of process vessels, alignments of motors, and driven units; damage to glass piping due to excessive vibrations.
- Check all pumps, agitators, venturi gas flow control valves, pH control systems, and confirm that adequate stock of neutralising chemicals are available.

For Pumps
- Check the suction line for possibility of having air pockets.
- Priming arrangements (dedicated overhead tank or connection from reliable source) should be provided for centrifugal pumps.

– High suction lift must be avoided as the pump will not get primed properly especially when the liquid temperature is more (due to vapour lock in the suction line).
– Foot valve flap operation should be smooth and strainer must be clean.
– Confirm that the piping on suction and discharge sides are fabricated correctly and are supported independently. They should not create tensile or compressive stress on the pump.
– Too thick/strong rupture disc shall not be provided for pressure release. Use discs of standard rating only.
– Check setting and operation of internal safety valve. It should be attended during erection of pump.
– Flow control valve and non-return valve shall be provided on discharge piping.

Instrumentation
• Avoid too short length of probes for temperature in a refractory lined unit. Provide protective sheaths for thermocouples.
• Provide strainers in sample lines for measuring pH, concentration of liquids.
• Confirm calibration of all flow and pH control systems, level indicators, thermocouples, warning alarms, links between field and control room instruments, etc.
• Check connecting cables and junction boxes. *Please refer to chapter on Process Control.*

Ensure Correct Levelling of
• Gas distributor and filter media in hot gas filters.
• Gas distributor and catalyst mass in fixed bed converter.
• Levelling of all equipment (base plates), storage tanks, day tanks, pumps, etc.
• Length and position of thermocouple probes in all process units and ducts.

9.10   Reducing Cost of Production

Some practical suggestions are given below for reducing cost of production.

9.10.1   Procurement of Raw Materials

Raw materials shall be procured from nearby suppliers. Contract may be made on a long-term basis for steady supply as per requirement. This will reduce the inventory (stock) of raw materials to be maintained at the plant warehouse. The cost of safe storage and premium for insurance will also reduce as a result.

Ensure that the raw materials are supplied as per specifications given by the production department. This can reduce the cost of pre-treatment operations such as crushing, pulverising, drying, filtration prior to feeding into the reaction units. This can ensure product quality which in turn minimises rejected quantities and operating problems.

- Procure the raw materials from reliable sources as per required quantities corresponding to amounts required for production (plus some margin) for preventing any interruption in supply due to any reason.
- It will be found that it can save costs of handling if it is possible to directly unload the raw material at point of feeding in the process. Example: procurement in pre-weighed bags of 50 kg each instead of carloads; procurement of solutions in carbuoys of small capacities of 10–20 L instead of procurement in tankers of 10–15 kl.
- If raw materials as per required specifications are not available at reasonable cost, then cheaper lower grade materials may be procured and purified by in-house treatment to improve their quality. However, the cost at feeding point of process units should be compared with higher cost of raw materials procured as per specifications.

9.10.2   Optimum Design, Operation, and Maintenance of Process Units and Machinery

- Check for any leaks of (treated) water, steam, fuel, compressed air, and attend immediately.
- Engage only well-qualified experienced manpower to avoid mistakes during operation. This can minimise mishaps; generation of effluents which in turn can reduce cost of liquid effluent treatment, air pollution control; and disposal of dangerous wastes. Provide refresher training at regular intervals to the personnel.
- Carefully monitor the plant operations and process parameters through process control facilities. It can reduce costs of maintenance of process units, machinery, and standby units
- Carefully maintain firefighting facilities, safety devices, and standby power supply units.
- Carry out regular research and development.
- Look into the administration expenses (transport, office supplies, welfare activities, etc.).
- Carry out timely payments of statutory taxes and operating fees for boilers, electrical systems, and storage of fuels to avoid penalties.
- These shall be able to deliver the required output when run at about 70--80% of their maximum capacity to avoid running at their peak capacity always. This can reduce wear and tear while running smoothly also.
- Do not use and operate a much bigger capacity reactor or furnace *instead of a smaller size required for the given production rates* since the bigger units can consume higher quantity of steam/fuel for heating as compared to one with smaller size.
- Install only sensitive and accurate controlling instruments for process control.

- All instrumentation must be regularly checked for accuracy and recalibrated immediately to ensure correct indication.
- Use only treated water for cooling towers, for making solutions for process, etc.
- Operate refrigeration system at optimum temperature to minimise power consumption. Avoid excessive cooling of process units if possible.
- Process tanks shall be provided with high level alarms and collection tanks to collect accidental overflows to prevent wastage of materials.
- Mechanical seals shall be provided for shafts of pumps and agitators handling dangerous fluids.
- Any leak of process material must be attended as soon as possible.
- Variable frequency drives shall be provided to drive motors of high capacity pumps and blowers to minimise power consumption.
- Minimise run time of high capacity equipment and motors.
- Use thermographic cameras to identify heat loss from process equipment, heat exchangers, steam lines, hot gas ducts, and attend the faults.
- Thermal insulation and cladding on process units, hot ducts, and pipelines must be well maintained.
- Optimise cost of catalyst against higher conversation efficiency and reduction in generation of effluents. Examine the cost of replacement of spares like filter cloth, active carbon, candle demisters, steam traps thermocouples, spray nozzles, heat exchanger tubes, and plates in filter press against the beneficial effect on plant efficiency.

9.10.3 Production Planning

- Avoid short production runs/frequent changeover of products since they need cleaning of process units and restart of reactors (which can require fuels, steam, electrical power during every change over)
- Maintain stocks of finished products in consultation with marketing (sales department) in order to minimise costs associated with excess inventories.
- This is to be done for continuous run of the plant rather than frequently stopping and starting again (since this can waste fuel and power during cooling down and requirement for reheating the process units again). The production planning is to be done by plant engineers in consultation with marketing department.
- Optimisation of batch sizes and different products when the same reactor and other connected process units are used. Frequent draining and cleaning the reactor can be avoided by producing the same product in repeated cycles. *It can cause some waste of reaction materials remaining inside (if different item is produced every time in consecutive batches).* There is also a chance of contamination of the products from previous batches of different products.

9.10.4   Reducing Maintenance Costs

- Well planned preventive maintenance programme is to be executed after consulting marketing, production, stores inventory control, and maintenance departments.
- Increase life of process units and machinery by standardisation of spares, simplification of design, strengthening of components subjected to more wear and tear; substitution by better (and cheaper if possible) components. This can reduce breakdowns, costs of frequently starting and stopping the plant; and funds required to maintain a larger inventory of different types of spares.
- Reduce separate stoppages for separate jobs... Carry out multiple maintenance jobs in a single stoppage as far as possible.
- Minor maintenance may be carried out during running of the equipment or during a short period of stoppage if it is safe to do so.

Major Maintenance/Replacement
- *This generally needs longer stoppage of running machinery and detailed planning for resources and the funds required. Senior engineers must look in to these.*
- Carefully compare the current loss of production *if the major maintenance is postponed* with the expected improvement *if maintenance is done immediately.*

9.10.5   Reducing Cost of Inventory of Process Spares

The inventory of spare parts and internals required by process units shall be reduced in order to minimise blocking of funds. This can be done by following methods:

- Standardisation of spares/internals like tower packings, candle demisters, filter cloth, separation screens, bags in bag filter, and catalyst support grids.
- Substitution by equipment/spares with better design and material of construction.
- Develop spares like valve keys, level indicators, and spray nozzles in house instead of purchasing from outside.
- Carry out reconditioning/repairs of steam traps, pump shafts, butterfly valves, catalyst support grids, etc. in own repair facilities.
- Minimising damage to process units by accurately calibrated and controlled feed systems for reactors, timely replacement of inactive catalyst, using only properly treated water for process solutions, maintaining clean heat transfer surfaces and cooling arrangements (wherever required), correct calibration of instruments and control systems, proper training of operating personnel, sufficient lighting of key areas, and controlling dust in working areas.
- Reuse of effluents in process at appropriate stage (to reduce consumption of raw materials by some recovery from the effluents). This will increase yield from the process plant.

- Recovery of products from effluents by recycling, purification such as distillation, crystallisation, and filtration, *if possible,* for minimising consumption of raw materials and utilities per unit output can also increase overall yield.

9.10.6 Cost of Sudden Shutdown

Any sudden shutdown of the running plant can result in wastage of raw materials (under process) remaining in various process units, feeding units, connecting ducts, and pipelines; can result in overflow from process vessels; release of gases from reactors or condensers; etc.

This can happen due to:

- The plant suddenly stops due to power outage.
- Due to sudden tripping of machinery by activation of safety interconnections provided.
- Hence provide:
  - DG sets which will automatically start in case of power outage.
  - Proper setting of safety interconnections.
  - Seal pots at end of gas release lines.
  - Non-return valves in feeding pipes to reactors.
  - Overhead cooling water tanks for supply to condensers.

This will enable quicker restart of the plant when power supply is resumed/when safety trips are reset. It can therefore reduce wastage of process materials.

Important Note *Experienced Operating and Maintenance personnel must be consulted during design, procurement, erection, and commissioning. They should be asked to examine these activities while being carried out. Their considered opinion should be discussed and implemented for safety and efficiency of the personnel and plant units.*

9.11 Environmental Pollution Control

9.11.1 Solid Waste

Solid wastes may contain significant amounts of valuable materials. If they are successfully reclaimed, recovered and reused, the volume of waste is considerably reduced. For example, the most promising reuse of bio waste is conversion of material to energy; recovery of sulphur from the sludge removed from melters.

9.11.2   Effluent Treatment Plant ETP

Following guidelines are for a typical ETP which can consist of an equalisation tank, pH adjustment system, polyelectrolyte dozer, lamella clarifier, tube settler, aeration unit, filter press, pressure sand filter, and active carbon filter.

Chemical engineer in charge of the operation of the plant shall check and monitor the following:

9.11.2.1   Wastewater Treatment

Purpose of wastewater treatment is to remove contaminants from water so that treated water can be reused/recycled in the plant to maximum extent possible and the discharge from the plant can be brought down to zero ZLD (please see Water Management in Chap. 1)

- Treatment process can be broadly classified as physical, chemical, and biological. Physical processes comprise screening, sedimentation, floatation, and filtration. Chemical processes include precipitation, coagulation, and disinfection. Biological process includes activated sludge process.
- Wastewater treatment processes are categorised as primary, secondary and tertiary.
- Primary treatment involves removal of identifiable suspended solids and floating matter. In secondary treatment organic matter that is in soluble or in colloidal form is removed. Tertiary treatment involves physical, chemical, or biological processes and used only when higher quality water is to be produced.

9.11.2.2   Operation of a Typical Effluent Treatment Plant

Monitor the physical and chemical analysis of the incoming effluent and its flow rate.

Confirm that the residence time of the effluents in the equalisation tank is about 24 h. There shall be slow agitation or (recycle if necessary) to make sure that the effluent at exit of the equalisation tank is having a uniform analysis. Total hold up of effluent in the tank should not normally exceed 80% of its full capacity. There shall be no short circuiting of the incoming and exit streams.

Remove settled sludge periodically from the equalisation tank.

pH control: Lime solution shall be always available for controlled addition to the effluent for neutralisation of any acidity. Monitor pH continuously by online instrumentation.

Flash mixer: This is to be run for thorough mixing of alum/ferrous sulphate to the neutralised effluent. It should have a propeller type high speed impeller. Sufficient stocks of alum/ferrous sulphate solution should be available always.

Polyelectrolyte dosing: This shall be done at a controlled rate through a rotameter by using a metering pump. The rate is to be adjusted for efficient settling.

Reaction chamber (after the above dosing system): shall have sufficient residence time for formation of floc. Confirm that there is no bypassing of the liquid from inlet to exit.

Lamella clarifier/Tube settler shall be regularly drained to remove accumulated sludge (the build-up of sludge at the bottom chamber should be monitored every 2 h at least).

If the exit liquid from the clarifier contains suspended particles the rate of effluent input to the treatment equipment units should be reduced.

The drained sludge shall be either centrifuged or passed through a filter press to squeeze out imbibed liquid and to remove the thickened sludge.

The sludge shall be sent to authorised secured land filling party.

Biological Treatment: Exiting liquid from the clarifier shall be aerated and treated by bacteria to further purify it. It is to be filtered while a part is to be recycled to maintain mother liquor suspended solids MLSS concentration. This must be monitored every hour.

Final Filtration

The filtrate is then generally to be passed through Pressure Sand Filter and Active Carbon Filter. Monitor the pressure drops and analysis of the finally exiting effluent.

Backwash/clean the units in case of high pressure drops/when the exit flow rate reduces considerably.

Recycle the exit effluent to tube settler or aeration tanks if the analysis does not meet given specifications.

9.11.2.3 Monitoring Other System Components

Monitor all pumps, air compressors, agitator assemblies, pH adjustment systems including the standby units regularly.

9.11.3 Air Pollution Control System

Typically, this consists of cyclone separator, impingement separators, bag filter, venturi scrubber, packed tower, ID fan, chimney, etc. Performance and condition of following units shall be monitored regularly during operation of the process plant and during maintenance.

Cyclone Separator, Impingement Separators
- Output of settled particulate matter from the unit.
- Concentration of particulate matter in the exit gas stream.
- Working of rotary air lock valve provided at the bottom side.
- Pressure drop across the unit.
- Any gas leak from the body.
- Check condition of the gas outlet pipe and support structure during maintenance.

Bag Filter
- Output of particulate matter drained out from the unit.
- Working of rotary air lock valve provided at the bottom side.
- Pressure drop across the unit. Very low pressure drop may indicate damaged bags.
- Any gas leak from the body.
- Working of pulse timer for loosening particulates deposited on the bags.
- Temperature of gases at inlet of the unit (shall not exceed limits given by manufacturer of bags to prevent damage).
- Monitor concentration of particulate matter in the exit gas stream regularly
- Check condition of the bags, pulse timer, and pulse air pipe during maintenance.

Venturi Scrubber
- Pressure drop across the unit (clean the pressure tap connections regularly).
- Working of throat control valve and external locking arrangement.
- Flow rate of scrubbing liquor (*please see* Sect. 9.5.3)
- Working of scrubbing liquor circulation pumps (including standby pump).
- Working of pH control system for scrubbing liquor.
- Confirm availability of sufficient stock of alkali.
- Any gas leak from the body.
- Analysis of incoming and exit gases to check efficiency of removal of particulate matter. The position of throat control valve is to be adjusted accordingly.
- Check condition of the throat control valve flap and operating mechanism during maintenance.

Packed Tower
- The diameter of the liquid outlet nozzle should be at least twice that of the liquid inlet nozzle. This is to minimise the possibility of holding up of scrubbing liquid in tower.
- Provide a view glass at a suitable place on the tower shell (to observe the incoming scrubbing liquor) if possible. A light glass shall also be provided diametrically opposite to the view glass.
- Monitor the pressure drop across the unit (clean the pressure tap connections regularly).
- Check working of scrubbing liquor spray nozzles.
- Working of scrubbing liquor circulation pumps (including standby pump). During initial start-up the flow of scrubbing liquid from the tower exit (return into the circulation tank) shall be observed.
- Flow rate of scrubbing liquor (*please see* Sect. 9.5.3) The level of liquid in the circulation tank shall also be checked every half an hour. It should remain constant.
- Working of pH control system for scrubbing liquor.
- Confirm availability of sufficient stock of alkali.
- Any leak of scrubbing liquor piping, gas inlet piping.
- Analysis of incoming and exit gases to check efficiency of removal of pollutants.
- The position of demister pad is to be checked and *reset if disturbed from position.*
- Check condition of demister pads/candles for any puncture during maintenance.

- Observe carefully presence of any damaged packing material in the exit liquid stream. *This may be observed by checking the circulation tank during any convenient stoppage of the unit.*

ID Fan
- Monitor the speed, current drawn by the drive motor and suction pressure.
- Monitor discharge pressure.
- Drain out any accumulated condensate from the body of the fan at regular intervals.
- Check foundation bolts, drive belts, and coupling.
- Attend immediately if any vibrations or abnormal noise is noticed.

Chimney
Check condition of:

- Observe the appearance of gases exiting from the chimney (smoky/blackish).
- Sampling system and connecting tubes.
- Lightening arrester.
- Aviation warning lamps.
- Drain valve at bottom (remove condensate regularly).
- Monkey ladder.
- Guy ropes (if not self-supported).
- *Check thoroughly if the chimney is getting tilted by itself. This can be a very rare thing however.*

9.12 Disposal of Dangerous CRIT Wastes

9.12.1 CRIT (Corrosive, Reactive, Inflammable, Toxic) Wastes

Many chemical industries generate wastes due to:

- Partially converted raw materials being taken out from the process reactors when they are cleaned before the next batch (batch production process) or during shutdown for maintenance in case of a continuous process plant. These materials will have to be disposed of if they are not useful for any purpose in the plant or cannot be sold to outside parties.
- Impurities separated during treatment/purification of raw materials before they are fed to the process units. Example: inert ash filtered out from liquid sulphur in a sulphuric acid plant. This ash also carries some sulphur with it which makes it combustible; and can release sulphur dioxide gases if it catches fire.
- Impurities separated from final product during purification.
- Waste/sludge removed during cleaning of process vessels, storage tanks, filters, heat exchangers, piping, and ducting; inactive catalyst; damaged spares, tower internals; effluent treatment units; etc. The quantities are generally considerable during annual overhaul of the plant when the process units are cleaned.

- These wastes may be corrosive, reactive, inflammable or toxic (CRIT) and cannot be disposed of as such without causing harm to the environment. Hence, they need to be treated further before final disposal.

9.12.2   Disposal Pathway

- Complete physical and chemical analysis of the waste is to be done to determine the safe methods of disposal whether by safe Incineration or by burial in a secure land fill site.
- Wastes having a calorific value CV of 2500 KCal/kg or high amounts of chlorides should be incinerated. There shall be an incinerator provided in the premises which is to be operated to burn the wastes in two stage incineration. The first stage is to be operated at 800 °C while the second stage is to be run at 1100 ° C with a residence time of 2 s at least. The exit gases are to be treated by a venturi scrubber followed by a packed tower with circulation of alkali solution in both.
- Disposal of waste with a CV less than 2000–2500 Kcal/kg or soluble chloride, nitrates, etc.: The waste is to be neutralised by lime powder and immobilised by inert ash (available from coal-fired thermal power stations) and portland cement. The impervious blocks of neutralised waste are then to be buried in a special land fill site specially constructed as per guidelines issued by statutory authorities for pollution control.
- The waste having higher chlorides, sulphates, nitrates, or a CV of more than 2500 Kcal/kg is to be destroyed in a suitable multistage facility. In modern practice this is done by plasma sasification vitrification reactors with multi stage air pollution control systems provided downstream.
- This method can destroy any type of waste while only a small quantity of impervious vitrified mass is produced which can be used for pavements (Fig. 9.1).

9.13   Water Treatment Plant

9.13.1   Following Shall Be Checked Every Shift

- Available stock of raw water and filtered water in their respective storage tanks.
- *Raw water source having more than 125 ppm TDS should not be accepted for use in the process. Try to obtain the supply of raw water from some other source if possible.*
- Available stock of treated water and its complete analysis especially TDS, pH, conductivity, dissolved oxygen, and silica content of the treated water.
- The stocks of treated water in their respective tanks must be as per specifications given by boiler manufacturer, as required for the process, as per specifications for make-up required for cooling towers, etc.

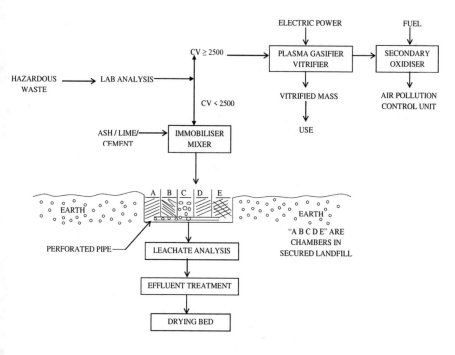

DISPOSAL OF HAZARDOUS WASTE

Fig. 9.1   Waste disposal

- HP boiler feed water should not contain more than 1.0 ppm TDS, more than 0.2 ppm of silica and more than 0.1 ppm of dissolved oxygen as it is detrimental to boiler tubes and to steam turbine for which HP steam is generally used.
- Process water should not generally have more than 50 ppm of total hardness.
- Soft water for LP boilers should not have more than 10 ppm total hardness and 10 ppm TDS.
- Check available stock of treatment chemicals like HCl, NaOH, NaCl, sodium meta bisulphite, $FeSO_4$, and polyelectrolyte.
- Check availability of spares for reverse osmosis plant as recommended by the supplier.
- Sufficient spare stock of resins should always be available for softener and DM plant resin columns.
- Continuously monitor recovery of condensate from all process units and its analysis, (pH, TDS, conductivity) since acidic condensate can corrode the boiler and economiser tubes. Investigate source of acidity and attend immediately.
- Collect all waste waters and treat for maximum reuse in the plant/premises (can be used for floor washing, gardening, fire-fighting, etc.).
- Analysis of treated water shall be monitored online at exit of softener, demineralisation unit, and RO units. This will enable to decide when resin columns of softeners/demineralisation plants are to be regenerated.

- Check conductivity of boiler feed water on-line continuously.
- Monitor working of degassing tower and de-aerator exit water temperature (higher temperature is useful for better removal of dissolved oxygen).
- Monitor the pressure drops in sand filter and active carbon filter.
- *High pressure drops indicate need for backwash.*
- Very low pressure drops indicate channelling of flow, possibility of damage to the support grids or internal spray nozzles, bypassing of the unit itself, etc.
- Regularly check the working of all pumps and blowers including standby units.
- Calibration of online instrumentation for ensuring quality of boiler feed water and feed water for other important process requirements should be confirmed regularly (at least once a week).

9.13.2 General

- All components of the reverse osmosis system, water pumps, instruments, and operating conditions must be monitored as per instructions from OEM of the system. Samples of exit water should be checked in laboratory also. Quantity of reject stream shall be monitored and efforts shall be made to use it in the premises elsewhere instead of discharging as effluent. *It can be distilled for reuse in the process if waste heat/exit steam from backpressure steam turbine is available in the plant. This will minimise the quantity of effluent.*
- The quantity of incoming water from all sources is to be measured every day. This is to be checked against the total process consumption; evaporation and windage losses from cooling towers; blow down from boilers, cooling towers, domestic/office consumption, etc. This includes monitoring of water consumption points like feed to process (for making solutions, for reactions, for combination as water of crystallisation, loss during evaporation, etc.); feed to boilers; as makeup to cooling towers; floor washing, gardening etc.
- Monitor volume of the exit stream of effluent from the premises. As per modern practice this should be zero. All efforts should be made to treat it and recycle the treated effluent so that draw of fresh water can be minimised. *Zero Liquid Discharge has become mandatory now.*
- Confirm recycling of condensate from all steam consuming (steam turbines) and heating units (evaporators, melters, reactor jacket) in the plant. Check analysis of the condensate—especially acidity and TDS—before recycling.
- The safety devices for RO plant and alarms for low level in tanks must be checked and confirmed to be working as per settings.

9.14 Electrical Installations and Equipment

Chemical engineers having responsibility of safe and efficient operation of the production plant should consult as well as advise electrical engineers on the following:-

- Location and installation of electrical equipment such as main metering panel, step-down and distribution transformers, rectifiers, motor control centre, routes of bus-bars and power cables, lighting network in the plant for ease of working in the plant.
- Total electrical load expected when the plant is run at rated capacity as per present product mix and for the future (planned) capacity after expansion or diversification.

9.14.1   Details of Facilities Which Will Need Power

- Construction power required for site fabrication of big process units and storage tanks (which are generally not transported from fabrication shops), big sized gas ducts, structural supports, assembly of chimney, etc.
- Site lighting for internal roads, fencing at periphery, and safe and secure storage of various items and plant machinery.
- Power required for erection work, mechanical trials, and commissioning of the plant.
- Power required for smooth operation of the plant at rated production capacity.
- There should be some margin for simultaneously running of standby machines (on trial through a bypass arrangement) along with running of main plant equipment.
- Estimate the maximum demand for power at present load conditions and for future requirements.
- Sanction is to be obtained accordingly from external power supply authority.

9.14.2   Electrical Load Details

Chemical engineers should inform the electrical engineers about details of the operating electrical load and assist in working out the load distribution in the plant:

- Load balancing in three phases.
- Estimation of AC and DC loads when plant is run at rated capacity.
- Additional power required when standby equipment are also run simultaneously (for trial runs) while the plant is also running.
- Margin for running the plant on an occasional 20% overload to meet urgent demand in the market; to meet unforeseen situation; and as a safety margin.
- Operating running time schedules /sequences of HT and LT Equipment (process units and plant machinery).
- Lighting in plant: indicate areas having inflammable vapours which should have flame proof lighting.

- Arrangement of alternate lights in working areas to be arranged on separate feeders.
- Rectifiers: Design of series and parallel arrangements of electrolytic cells as per individual cell design and overall production rate desired.
- Operating sequence of the electrolytic cells.
- Requirement of variable frequency drives for reducing power consumption by large capacity pumps and blowers (for avoiding unnecessary running at higher speeds when lower flow rates are required).
- Earth connections for storage tanks, process units, and piping for handling inflammable materials. These connections shall be regularly checked for continuity.
- Flame proof motors for agitators fitted on reactors handling inflammable materials.
- High torque motorised valves for process control and for safety purpose with remote operation control.
- Planning for cogeneration with synchronous generator/induction generator as per in-house steam generation, and requirements of power and steam in the plant.
- Study the steam turbo generator set capacity and external supply reliability.
- Diesel generator sets--planning for capacity required with features of auto start on mains failure (and auto load sharing amongst more than one DG set if possible).
- Emergency power supply planning for certain important drives for boiler feed water pumps, process safety, firefighting, pollution control, etc.
- Provision of power supply for cathodic protection of plant units. A list of such units shall be made and given to electrical engineers.
- Monitoring of maximum draw of power during plant operations.
- Protection facilities for motors and other equipment.
- Indoor and outdoor installation of equipment.
- Condition monitoring of electrical equipment.
- Provision of in-house facilities for maintenance.
- Programme for major maintenance and annual shutdown when the electrical load will be much less than normal load.
- Requirement of power during annual shutdown and subsequent start-up should be informed to the electrical engineers.

For further information on Waste Disposal the readers may contact SMS Infrastructure Ltd, IT Park Road, Shreenagar Parsodi, Nagpur 440022, India

References/for Further Reading

1. "Process Equipment Procurement in the Chemical and Related Industries" Kiran Golwalkar, Published by Springer International Publishing, Switzerland.
2. "Production Management of Chemical Industries" by Kiran Golwalkar, Published by Springer International Publishing, Switzerland.

Appendix A: Material of Construction

Physical Properties of Various Materials of Construction

Hastelloy C—276

Sp.Gr. 8.89
Specific Heat: 427 J/kg.K
Melting Point: 1323–1371 °C
Thermal Conductivity: 10.5 W/m.K (at 50 °C)
Applications: Conc. sulphuric acid service

Titanium

Sp.Gr. 4.5
Specific Heat: 540 J/kg.K
Melting Point: 1668 °C
Thermal Conductivity: 21.9 W/m·K (at 21 °C)
Applications: Chlor-alkali industry, petrochemical fibre, filter, heat exchanger, condenser

Tantalum

Sp.Gr. 16.6
Specific Heat: 140 J/kg.K
Melting Point: 3017 °C
Thermal Conductivity: 57.5 W/(m·K)
Applications: Shell and tube heat exchangers, capacitors

Monel—400

Sp.Gr. 8.80
Specific Heat: 427 J/kg.K(at 21 °C)
Melting Point: 1300–1350 °C

© The Author(s), under exclusive license to Springer Nature Switzerland AG 2022   253
K. R. Golwalkar, R. Kumar, *Practical Guidelines for the Chemical Industry*,
https://doi.org/10.1007/978-3-030-96581-5

Thermal Conductivity: 22 W/m.K(at 21 °C)

Applications: Hydrocarbon processing equipment, valves, pumps, shafts, fittings, and heat exchangers

Copper

Sp.Gr. 8.96

Specific Heat: 385 J/kg.K(at 25 °C)

Melting Point: 1084.62 °C

Thermal Conductivity: 401 W/(m·K)

Applications: Wiring, motors, utensils, coils.

Admiralty Brass

Sp.Gr. 8.53

Specific Heat: 380 J/kg.K

Melting Point: 900–940 °C

Thermal Conductivity: 111 W/m.K

Applications: Condenser tubes, evaporator and heat exchanger tubes, steam condensers

70—30 Copper-Nickel

Sp.Gr. 8.94

Specific Heat: 377 J/kg.K

Melting Point: 1171 °C

Thermal Conductivity: 29 W/(m·K)

Applications: Salt water flanges, pump impellers, valve bodies

Carbon Steel AISI 1020

Sp.Gr. 7.87

Specific Heat: 486 J/kg.K

Melting Point: 1515 °C

Thermal Conductivity: 51.9 W/mK

Applications: Shafts, chains, lightly stressed gears

Gray Cast Iron

Sp.Gr. 7.5

Specific Heat: 490 J/kg-K

Melting Point: 1390 °C

Thermal Conductivity: 41 W/m-K

Applications: Pump housings, valve bodies, electrical boxes,

Stainless Steels AISI 410

Sp.Gr. 7.8

Specific Heat: 460 J/kg-K

Melting Point: 1495 °C

Thermal Conductivity: 24.9 W/m.K (at 100 °C)

Applications: Pipelines, valves, and nozzles

Stainless Steel AISI 446

Sp.Gr. 7.7
Specific Heat: 490 J/kg-K
Melting Point: 1510 °C
Thermal Conductivity: 17 W/m-K
Applications: Burners and boiler parts

Stainless Steel AISI 304

Sp.Gr. 7.9
Specific Heat: 490 J/kg-K
Melting Point: 1400 °C
Thermal Conductivity: 14 W/m-K
Applications: Fasteners, heat exchangers, chemical containers

Stainless Steel AISI 310

Sp.Gr. 7.8
Specific Heat: 490 J/kg-K
Melting Point: 1454 °C
Thermal Conductivity: 11 W/m-K
Applications: Kilns, radiant tubes, tube hanger, fluidised bed combustors

Nickel Alloy 400

Sp.Gr. 8.9
Specific Heat: 427 J/kg-K
Melting Point: 1300–1350 °C
Thermal Conductivity: 22 W/m-K
Applications: Seawater desalination plants

Nickel—Molybdenum alloy B-2

Sp.Gr. 9.3
Specific Heat: 390 J/kg-K
Melting Point: 1570 °C
Thermal Conductivity: 11 W/m-K
Applications: Pumps, valves, mechanical seals, rupture discs, flanges

Tin

Sp.Gr. 7.31
Specific Heat: 210 J/kg.K
Melting Point: 231.9 °C
Thermal Conductivity: 63.2 W/mK
Applications: Solder, bearings, cans

Zirconium

Sp.Gr. 6.52
Specific Heat: 270 J/kg.K
Melting Point: 1855 °C
Thermal Conductivity: 22.6 W/(m·K)
Applications: Steel alloys, colored glazes, bricks, ceramics, abrasives (Table A.1).

Note: Plant engineers are requested to refer to Engineering Tool Box and Chemical Engineers Handbook for more information

• The authors of the present book gratefully thank the editor authorities of Engineering Tool Box for making available the above information on the internet and acknowledge the source of their information as Engineering Tool Box. Engineering Tool Box, (2003). *Metals and Corrosion Resistance*. [online] Available at: https://www.engineeringtoolbox.com/metal-corrosion-resistance-d_491.html.

Table A.1 Corrosion resistance of commonly used materials for various fluids

Fluid	Carbon Steel	Cast Iron	302 and 304 Stainless Steel	316 Stainless Steel	Bronze	Durimet	Monel	Hasteloy B	Hasteloy C
Acetic acid, aerated									
Acetic acid, vapors								Na	
Acetone									
Alcohols									
Ammonia									
Ammonium chloride									
Ammonium Nitrate									
Aniline									
Benzene (benzol)									
Calcium Chloride (alkaline)									
Carbon dioxide, dry									
Carbon dioxide, wet									
Chlorine gas									
Chlorine gas, wet									
Chlorine, liquid									
Chromic acid									
Ethane									
Ethylene									
Ferric chloride									
Formaldehyde									
Formic acid									
Gasoline									
Hydrochloric acid, aerated									
Hydrochloric acid, air free									
Hydrogen									
Hydrogen peroxide	Na								
Hydrofluoric acid, aerated									
Hydrogen sulfide, moist									
Natural gas									
Nitric acid									
Oxygen									
Petroleum oils									
Sodium chloride									
Sodium hydroxide									
Sodium Hypochlorite									
Sulfur dioxide, dry									
Sulfur trioxide, dry									
Sulfuric acid, aerated									
Sulfuric acid, air free									
Water, steam boiler feeding system									
Water, distilled									
Water, sea									

Green…. Can be used
Yellow…. Use with caution
Red …. Not suitable
na—data not available

Appendix B: Pressure Vessels (Table B.1)

Table B.1 ASME/ANSI codes for pressure vessels

Code section	Description		
ASME Section 1: Boilers and component parts	Rules in ASME's Boiler and Pressure Vessel Code (BPVC) are applicable to boilers and component parts including piping constructed. These rules apply to boilers in which steam or other vapor is generated at pressures exceeding 15 psig, and high temperature water boilers intended for operation at pressures exceeding 160 psig and/or temperatures exceeding 250 °F. Superheaters, economizers, and other pressure parts connected directly to the boiler without intervening valves are considered as part of the scope of Section I		
ASME Section II: Material specifications	Subpart A—Ferrous material specifications		
	Subpart B—Non-ferrous material specifications-materials		
	Subpart C—Specifications for welding rods, electrodes, and filler metals		
	Subpart D—properties mentioned in three subparts with tables for: Stress, physical properties, determining shell thickness of components under external pressure		
ASME Section III: Rules for construction of nuclear facility components	Division 1: Rules for construction of nuclear facility components	Subsection NB: Class 1 components	
		Subsection NC: Class 2 components	
		Subsection ND: Class 3 components	
		Subsection NE: Class MC components	
		Subsection NF: Supports	
		Subsection NG: Core support structures	
	Division 2: Code for concrete reactor vessels and containment		
	Division3: Containment systems for storage and transport packaging of spent nuclear fuel and high level radioactive material and waste		

(continued)

Table B.1 (continued)

Code section	Description	
ASME Section IV: Rules for construction of heating boilers	The rules apply to steam heating boilers and hot water heating boilers.	
	These rules are used for safe design, construction, installation, and inspection of boilers.	
ASME Section V: Non-destructive examination	Contains requirements and methods for non-destructive examination which are Code requirements to the extent they are specifically referenced and required by other Code Sections	
	These non-destructive examination methods are intended to detect surface and internal discontinuities in materials, welds, and fabricated parts and components	
	They include radiographic examination, ultrasonic examination, liquid penetrant examination, magnetic particle examination, eddy current examination, visual examination, leak testing, and acoustic emission examination	
ASME Section VI: Recommended rules for the care and operation of heating boilers	General, covers scope and terminology	
	Types of boilers	
	Accessories and installation	
	Fuels	
	Fuel burning equipment and fuel burning controls	
	Boiler room facilities	
	Operation, maintenance, and repair-steam boilers	
	Operation, maintenance, and repair of hot water boilers and hot water heating boilers	
	Water treatment	
ASME Section VII: Recommended guidelines for the care of power boiler	Guidelines to promote safety to power boilers that produce steam for external use at a pressure exceeding 15 psig	
ASME Section VIII: Rules for construction of pressure vessels	Division 1	– Subsection A is general pressure vessel information – Subsection B covers the requirements pertaining to methods of fabrication of pressure vessels – Subsection C lists the requirements pertaining to classes of materials
	Division 2	Alternative rules for construction of pressure vessels
	Division 3	Alternative rules for construction of high pressure boilers

Code section	Description
ASME Section IX: Welding and brazing qualifications	Covers the requirements for Weld Procedure Specifications (WPS), Procedure Qualification Records (PQR), and certification requirements for tackers, welders, welding operators, and brazing personnel
ASME Section X: Fiber-reinforced plastic pressure vessels	Provides requirements for construction, fabrication, inspection and testing methods required for FRP pressure vessel

Appendix C: Piping and Pumping Systems (Tables C.1, C.2, and C.3)

Readers may refer to Piping colour code IS 2379 for all chemicals

Piping Sample Calculations

Sample Problem 1

Develop a family of curves of pressure drop along standard steel pipe of length 300 m for pipe diameters of 400 mm, 450 mm, and 500 mm for different flowrates of 0.4 m³/h and 0.5 m³/h and 0.6 m³/h, considering fluid flowing as water (Fig. C.1).

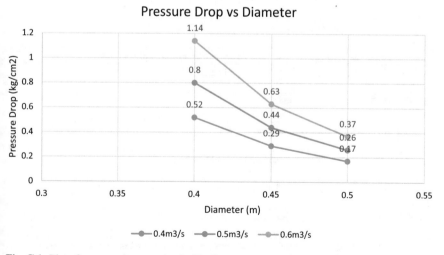

Fig. C.1 Plot of pressure drop vs. pipe inside diameters

Table C.1 ASME/ANSI codes for piping

Code	Description
ASME B31.1	Power piping: Provides rules for piping typically found in electric power generating stations, in industrial and institutional plants, geothermal heating systems, and central and district heating and cooling systems
ASME B31.2	Fuel gas piping
ASME B31.3	Process piping
ASME B31.4	Liquid transportation systems for hydrocarbons, liquid petroleum gas, hydrocarbons, anhydrous ammonia, alcohol and other liquids
ASME B31.5	Piping refrigeration and heat transfer components
ASME B31.8	Gas transmission and distribution piping systems
ASME B31.9	Building services piping
ASME B31.11	Slurry transportation piping systems
ASME B 31.3	Chemical plant and petroleum refinery piping
ANSI/ASME A13.1	Scheme for the identification of piping system
ANSI/ASME B 16.1	Cast iron pipe flanges and flanged fittings
ANSI/ASME B 16.3	Malleable-iron threaded fitting
ANSI/ASME B 16.4	Grey iron threaded fittings
ANSI/ASME B 16.5	Pipe flanges and flanged fittings
ANSI/ASME B 16.9	Factory-made wrought steel butt welding
ANSI/ASME B 16.10	Face to face and end to end dimensions of valves
ANSI/ASME B 16.11	Forged fitting, socket welding and threaded
ANSI B 16.20	Metallic gaskets for pipe flanges—Ring joint, spiral wound
ANSI/ASME B 16.21	Non metallic gaskets for pipe flanges
ANSI/ASME B 16.25	Butt welding ends
ANSI/ASME B 16.28	Short radius elbows and returns
ANSI/ASME B 16.34	Steel valves, flanged and butt welding ends
ANSI/ASME B 16.42	Ductile iron pipe flanges and flanged fittings—Class 150# and 300#
ANSI/ASME B 16.47	Large diameter steel flanges—NPS—26″ to 60″
ANSI/ASME B 18.2 1 and 2	Square and hexagonal head bolts and nuts (inch and mm) ANSI/ASME B 36.10—Welded and seamless wrought steel pipes
ANSI/ASME B 36.19	Welded and seamless austenitic stainless steel pipe

Table C.2 API codes for pipes

API code	Description
API 5L	Specification for line pipe
API 6D	Pipeline valves, end closures, connectors and swivels
API 6F	Recommended practice for fire test for valves
API 593	Ductile iron plug valves—Flanged ends
API 598	Valve inspection and test
API 600	Steel gate valves
API 601	Metallic gaskets for refinery piping
API 602	Compact design carbon steel gate valves
API 604	Ductile iron gate valves—Flanged ends
API 605	Large diameter carbon steel flanges
API 607	Fire test for soft seated ball valves
API 609	Butterfly valves
API 1104	Standard for welding pipeline and facilities

Table C.3 Piping colour code as per IS 2379 for few chemicals

Flow medium	Ground colour	First colour band	Second colour band	Third colour band
1. *Water and steam*				
Drinking water	Green	Blue	Red	–
Make up water	Green	–	–	–
Waste water	Black	–	–	–
Cooling water supply	Green	Blue	–	–
Cooling water return	Green	Brown	–	–
Boiler feed water	Green	Orange	–	–
Condensate	Green	Grey	–	–
Steam	Grey	–	–	–
2. *Air and gases*				
Compressed air	Blue	–	–	–
Blast furnace (cold)	Aluminum	French blue	Golden yellow	French blue
Blast furnace(hot blast)	Aluminum	–	–	–
Vacuum	Blue	Black	–	–
Mixed gases	Aluminium	Golden yellow	Signal red	Golden yellow
Blast furnace gas (clean)	Aluminium	Golden yellow	Black	Golden yellow
Blast furnace gas (raw)	Aluminium	Golden yellow	Brilliant green	Golden yellow
Synthesis gas	Aluminium	Golden yellow	Signal red	–
Ammonia gas	Aluminium	Golden brown	–	–
Nitrogen	Yellow	Black	–	–
Oxygen	Yellow	White	–	–
Acetylene	Yellow	Violet	–	–
3. *Oil*				

(continued)

Table C.3 (continued)

Flow medium	Ground colour	First colour band	Second colour band	Third colour band
Lubricating oil	Brown	Grey	–	–
Hydraulic oil	Brown	Violet	–	–
Furnace oil	Brown	Red	–	–
Ethylene glycol	Dark Admirality grey	Brilliant green	Gulf red	
Ethylene Di-chloride	Dark Admirality grey	Gulf red		
Benzene	Dark Admirality grey	Canary yellow		
Butadine	Dark Admirality grey	Black		
Acetone	Dark Admirality grey	Black	Canary yellow	
Methanol	Dark Admirality grey	Deep buff		
Naptha	Dark Admirality grey	Light brown	Black	
4. ACIDS				
Phospdoric acid	Dark violet	Silver grey		
Hydrofluorie acid	Dark violet	Signal red	French blue	
Sulphuric acid	Dark violet	Brilliant green	Light orange	
Nitric acid	Dark violet	French blue	Light orange	
Hydrochloric acid	Dark violet	Signal red	Light orange	
Acetic acid	Dark violet	Silver grey		
5. Chemical and allied products				
Brine	Black	White		
Caustic solution	Smoke grey	Light orange		
Lime	Smoke grey	White	Canary yellow	
Carbon disulphide	Black	Light orange		
Strong caustic	Smoke grey	French blue	White	
Sodium carbonate solution	Dark violet	Jasmine yellow		
Ammonia	Canary yellow	Dark violet		
Chlorine	Canary yellow	Dark violet	Light orange	
Sulphur dioxide	Canary yellow	Dark violet	Golden brown	
Acetylene	Canary yellow	Service brown		
Flare gases	Canary yellow			
Hydrogen sulphide	Canary yellow	Gulf red		
Blast furnace gas	Canary yellow	Signal red	Light grey	
Butane	Canary yellow	Signal red		
Coal gas	Canary yellow	Signal red	Brilliant green	

(continued)

Flow medium	Ground colour	First colour band	Second colour band	Third colour band
Carbon dioxide	Canary yellow	Light grey		
Ethylene	Canary yellow	Dark violet	Signal red	
Ethylene oxide	Canary yellow	Dark violet	Brilliant green	
Hydrogen	Canary yellow	Signal red	French blue	
Methane	Canary yellow	Signal red	Light brown	
Nitrogen	Canary yellow	Black		
Oxygen	Canary yellow	White		
Propane	Canary yellow	Signal red	Black	
Phosgene	Canary yellow	Black	White	

Sample calculations:
We know that pressure drop due to skin friction in pipes is given by

$$\frac{\Delta p}{g} = f \times \frac{L}{D} \times \frac{v^2}{2g} \times \rho$$

where, Δp = Pressure drop; L = Length of pipe; D = Inside diameter of pipe; V = Velocity of flow; ρ = Density of fluid; F = Friction factor.

$$\text{Area of flow} = A = \pi r^2 = \pi (0.4/2)^2 = 0.1256 \text{m}^2$$

$$\text{Velocity} = V = \text{Flow rate / area} = 0.4 / \times 0.1256 = 3.185 \text{m/s}$$

$$\text{Reynolds number} = \rho V D / \mu = (1000 \times 3.185 \times 0.4) / 0.001 = 1273239.5$$

$$\text{Surface roughness for steel pipe} = k = 0.045$$

$$\text{Relative roughness} = k / D = 0.045 / 400 = 1.124^* 10^{-4}$$

Friction factor, $f = 0.0137$ (By Colebrook relation or Moody's Figure)
Pressure drop, Δp from the relation, $\frac{\Delta p}{g} = f \times \frac{L}{D} \times \frac{v^2}{2g} \times \rho$

$$\Delta p = 0.5208 \text{kg/cm}^2$$

Calculation table: (Values taken from PYTHON CODE OUTPUT below):
Pressure Drop for flow rate 0.4 m³/s (1440 m³/h)

Diameter (m)	Area (m²)	Velocity (m/s)	Reynold's number	Relative roughness	Friction factor	Pressure drop (kg/cm²)
0.4	0.1256	3.185	1273239.5	0.0001124	0.0137	0.5208
0.45	0.159	2.516	1131768.4	9.99e-05	0.0136	0.2874
0.5	0.19625	2.038	1018591.6	8.99e-05	0.0136	0.1692

Pressure Dropfor flow rate 0.5 m³/s (1800 m³/h)

Diameter (m)	Area (m²)	Velocity (m/s)	Reynold's number	Relative roughness	Friction factor	Pressure drop (kg/cm²)
0.4	0.1256	3.981	1591549.4	0.0001124	0.0135	0.802239
0.45	0.1589625	3.1454	1414710.6	0.000099	0.0134	0.441839
0.5	0.19625	2.5477	1273239.5	0.000089	0.0133	0.259668

Pressure Drop for flow rate 0.6 m³/s (2160 m³/h)

Diameter (m)	Area (m²)	Velocity (m/s)	Reynold's number	Relative roughness	Friction factor	Pressure drop (kg/cm²)
0.4	0.1256	4.777	1909859.3	0.0001124	0.0134	1.14340
0.45	0.1589625	3.7745	1697652.7	0.000099	0.0133	0.62873
0.5	0.19625	3.0573	1527887.4	0.000089	0.0132	0.36895

Python Code for Sample Problem 1

Program:

```python
import numpy as np
A = [0, 0, 0]; k = [0, 0, 0]; v = [0, 0, 0]; Re = [0, 0, 0]
Qhr = float(input("Enter the value of Vol Flow Rate of fluid in pipe
(m3/h): "))
Q = Qhr / 3600
L = float(input("Enter the value of length of the pipe (m): "))
rho = float(input("Enter the value of density of fluid (kg/m3): "))
eps = float(input("Enter the value of roughness factor of pipe mate-
rial (mm): "))
eps = eps*10**-3
mu = float(input("Enter the value of viscosity of fluid (Pa-s): "))
g = 9.81
D = [0.4, 0.45, 0.5]
for i in range(3):
 A[i] = np.pi * (D[i]/2)**2
 k[i] = eps / D[i]
 v[i] = Q / A[i]
 Re[i] = (D[i] * v[i] * rho) / mu

'''def f(Re, eps, D):
 f = 8 * ( (8 / Re)**12 + ((37530 / Re)**16 + (-2.457 * np.log((7
/ Re)**0.9 + 0.27*(eps / D)))**16)**-1.5)**(1/12)
 return f'''

print("The Reynold's Number is", Re, "and the Relative Roughness
is", k, ".")
f = [0, 0, 0]; h = [0, 0, 0]; delpbyg = [0, 0, 0]
for i in range(3):
 f[i] = 8 * ( (8 / Re[i])**12 + ((37530 / Re[i])**16 + (-2.457 *
np.log((7 / Re[i])**0.9 + 0.27*(eps / D[i])))**16)**-1.5)**(1/12)
 h[i] = (f[i] * L * v[i]**2) / (2 * g * D[i])
 delpbyg[i] = h[i] * rho
for i in range(3):
 print("The value of Pressure drop per unit gravitational force
acting is:", delpbyg[i]*10**-4, "kg/cm2")

Output (1440 m3/h or 0.4 m3/s):
Enter the value of Vol Flow Rate of fluid in pipe (m3/h): 1440
Enter the value of length of the pipe (m): 300
Enter the value of density of fluid (kg/m3): 1000
Enter the value of roughness factor of pipe material (mm): 0.045
Enter the value of viscosity of fluid (Pa-s): 0.001
The Reynold's Number is [1273239.5447351628, 1131768.4842090334,
1018591.6357881301]     and     the     Relative     Roughness     is
[0.00011249999999999998,                    9.999999999999999e-05,
```

8.999999999999999e-05] .
The value of Pressure drop per unit gravitational force acting is:
0.5208298214101177 kg/cm2
The value of Pressure drop per unit gravitational force acting is:
0.28744543827110863 kg/cm2
The value of Pressure drop per unit gravitational force acting is:
0.16925856046401544 kg/cm2
Process finished with exit code 0

Output (1800 m³/h or 0.5 m³/s):
Enter the value of Vol Flow Rate of fluid in pipe (m3/h): 1800
Enter the value of length of the pipe (m): 300
Enter the value of density of fluid (kg/m3): 1000
Enter the value of roughness factor of pipe material (mm): 0.045
Enter the value of viscosity of fluid (Pa-s): 0.001
The Reynold's Number is [1591549.4309189534, 1414710.6052612918,
1273239.5447351628]   and   the   Relative   Roughness   is
[0.00011249999999999998,   9.999999999999999e-05,
8.999999999999999e-05] .
The value of Pressure drop per unit gravitational force acting is:
0.8022393311489164 kg/cm2
The value of Pressure drop per unit gravitational force acting is:
0.44183932458043457 kg/cm2
The value of Pressure drop per unit gravitational force acting is:
0.25966822406872203 kg/cm2
Process finished with exit code 0

Output (2160 m³/h or 0.6 m³/s):
Enter the value of Vol Flow Rate of fluid in pipe (m3/h): 2160
Enter the value of length of the pipe (m): 300
Enter the value of density of fluid (kg/m3): 1000
Enter the value of roughness factor of pipe material (mm): 0.045
Enter the value of viscosity of fluid (Pa-s): 0.001
The Reynold's Number is [1909859.317102744, 1697652.72631355,
1527887.4536821952]   and   the   Relative   Roughness   is
[0.00011249999999999998,   9.999999999999999e-05,
8.999999999999999e-05].
The value of Pressure drop per unit gravitational force acting is:
1.1434027274049248 kg/cm2
The value of Pressure drop per unit gravitational force acting is:
0.6287374131131898 kg/cm2
The value of Pressure drop per unit gravitational force acting is:
0.3689501523200388 kg/cm2

NOTE: Above code can be used to determine pressure drop for varying flow rates, fluids, and pipe lengths.

Sample Problem 2

Estimate power required for a pump for transferring water from one tank to another located at height of 20 m. Water has to be transferred at a rate of 35 m³/h (Fig. C.2).

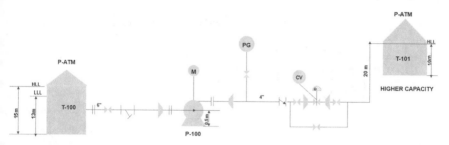

Fig. C.2 Schematic of sample problem 2 for pump power calculation

Process Data

Process fluid:	Water
Fluid temperature (T):	40 °C
High suction liquid level above grade:	15 m
Low suction liquid level above grade:	13 m
Impeller centreline level above grade:	0.5 m
Pipe roughness (carbon steel):	0.047 mm
Pump capacity: (Normal)	35 m³/h
Vapour pressure	1.086 kg/cm²a
Density	1000 kg/m³
Viscosity	1 cP
Tube inner diameter on suction side	154.06 mm
Tube inner diameter on discharge side	102.26 mm
Strainer pressure drop	0.05 kg/cm² on suction side
Gate valves	3 on suction side and 2 on discharge side
Check valves	1 on discharge side
Globe valve	1 on discharge side
STD 90° elbows	3 on suction side and 4 on discharge side.
At rated capacity control valve pressure drop	2 kg/cm² on discharge side
Pipe straight length on suction side	6 m
Pipe straight length on discharge side	23 m
Discharge static head	20 m
Low suction liquid level w.r.t grade in meters	13 m

Calculation: (Values taken from PYTHON CODE OUTPUT below)
Design flow rate suction and discharge

$$Q = Q_n \left[1 + \frac{d_m}{100} \right]$$

$$= 35 \left[1 + \frac{10}{100} \right]$$

$Q = 38.5 \text{ m}^3\text{/h}$

Q_n = normal flow rate
d_m = design margin flow
Q = design flow rate

Suction line velocity (V_s)

$$V_s = \left[\frac{Q}{3600} \right] \left[\frac{1}{\frac{\pi}{4}\left(\frac{\text{IDsuct}}{1000} \right)^2} \right]$$

$$= \left[\frac{38.5}{3600} \right] \left[\frac{1}{\frac{\pi}{4}\left(\frac{154.06}{1000} \right)^2} \right]$$

$V_s = 0.5737 \text{ m/s}$

ID suct = tube inner diameter suction side

Discharge line velocity (V_d).

$$V_d = \left[\frac{Q}{3600} \right] \left[\frac{1}{\frac{\pi}{4}\left(\frac{\text{ID}_{disc}}{1000} \right)^2} \right]$$

$$= \left[\frac{38.5}{3600} \right] \left[\frac{1}{\frac{\pi}{4}\left(\frac{102.26}{1000} \right)^2} \right]$$

$V_d = 1.3021 \text{ m/s}$

ID disc = tube inner diameter discharge side

Reynold's number suction line

$$N_{ReS} = \frac{\left[\frac{\text{ID}_{suc}}{1000} \right] \cdot V_s \cdot \rho \cdot 1000}{\mu}$$

$$= \frac{\left[\frac{154.06}{1000} \right] \cdot 0.5737 \cdot 1000 \cdot 1000}{\mu}$$

$$= 88384.97$$

Reynold's number discharge line

$$N_{ReD} = \frac{\left[\dfrac{ID_{dis}}{1000}\right] \cdot V_d \cdot \rho \cdot 1000}{\mu}$$

$$= \frac{\left[\dfrac{102.26}{1000}\right] \cdot 1.3021 \cdot 1000 \cdot 1000}{1}$$

$$= 133156.55$$

Friction factor (Churchill's equation) for suction line:

$$f_s = 8 * \left[\left(\frac{8}{N_{Re}}\right)^{12} + \left[\left(\frac{37530}{N_{Re}}\right)^{16} + \left[-2.457 * \ln\left\{\left(\frac{7}{N_{Re}}\right)^{0.9} + 0.27\left(\frac{\varepsilon}{ID}\right)\right\}\right]^{16}\right]^{-1.5}\right]^{\frac{1}{12}}$$

$$f_s = 0.0199$$

Friction factor (Churchill's equation) for discharge line:

$$f_d = 8 * \left[\left(\frac{8}{N_{Re}}\right)^{12} + \left[\left(\frac{37530}{N_{Re}}\right)^{16} + \left[-2.457 * \ln\left\{\left(\frac{7}{N_{Re}}\right)^{0.9} + 0.27\left(\frac{\varepsilon}{ID}\right)\right\}\right]^{16}\right]^{-1.5}\right]^{\frac{1}{12}}$$

$$f_d = 0.0195$$

Calculation of equivalent Le/D for suction side.

Item	Le/D	Qty	Total Le/D
Gate valve	13	3	39
90° elbow	30	3	90
Total			129

Total Equivalent length

$$Le / D = 129$$

$$Le = 129 \times 0.15406$$
$$= 19.87\text{m}$$

$$Total\,Le = \text{Straight length} + \text{Equivalent length}$$
$$= 6 + 19.87$$
$$= 25.87\text{m}$$

Pressure drop in suction line in kg/cm^2

$$\Delta P_{suc} = \frac{\rho^* f_s^* \text{Le}_{suc}^* V_s^2}{2^* 9.81^* \text{ID}_{suc}^* 10}$$

$$= \frac{1000^* 0.0199^* 25.87^* (0.5737)^2}{2^* 9.81^* 154.06^* 10} = 0.0056 \text{Kg} / \text{cm}^2$$

Taking strainer pressure drop = 0.05 kg/cm^2

Total pressure drop at suction, $\Delta P_{\text{total suct}}$ in kg/cm^2

$$\Delta P_{\text{total suct}} = \Delta P_{\text{suct}} + \text{Strainer pressure drop}$$

$$= 0.0056 + 0.05$$

$$= 0.0556 \text{Kg} / \text{cm}^2$$

Calculation for Equivalent length for discharge line.

Item	Le/D	Qty	Total Le/D
Gate valve	13	2	26
90° elbow	30	4	120
Check valve	135	1	135
Globe valve	450	1	450
Total			731

Total Equivalent length

Le / D = 731

$$\therefore \text{Le} = 731 \times 0.10226$$

$$= 74.75 \text{ m}$$

$$\text{Total Le} = \text{Straight length} + \text{Equivalent length}$$

$$= 23 + 74.75$$

$$= 97.75 \text{m}$$

Pressure Drop in Discharge Line in kg/cm^2

$$\Delta P_{dis} = \frac{\rho^* f_d^* \text{Le}_{dis}^* V_d^2}{2^* 9.81^* \text{ID}_{dis}^* 10}$$

$$= \frac{1000^* 0.0195^* 97.75^* (1.3021)^2}{2^* 9.81^* 102.26^* 10}$$

$$= 0.1610 \text{Kg} / \text{cm}^2$$

Compounding with other pressure drops:

$$Z_d = 20 \text{m}$$

where, Z_d is Discharge static head in meter

$$\Delta P_{\text{Control Valve}} = 2\,\text{kg}/\text{cm}^2$$

$$\Delta P_{\text{totaldis}} = \Delta P_{\text{dis}} + \Delta P_{\text{Control Valve}} + \frac{Z_d^{\,*}\text{Specific gravity}}{10}$$

$$= 0.1610 + 2 + \frac{20^*1}{10} = 4.161\,\text{Kg}/\text{cm}^2$$

Suction Pressure, P_{suc}

$$P_{\text{suc}} = \left[P_{\text{suct}} + \left(Z_{\text{sl}} - Z_{\text{pump}} \right) * \frac{\text{Specific Gravity}}{10} - \Delta P_{\text{totalsuc.}} \right]$$

$$= \left[1.033 + \left(13 - 0.5 \right) * \frac{1}{10} - 0.0556 \right] = 2.2274\,\text{Kg}/\text{cm}^2$$

where, P_{suct} = Pressure at liquid surface Kg/cm²; Z_{sl} = Low suction liquid level w.r.t grade in meters; Z_{pump} = Pump suction nozzle level w.r.t grade in meters; $\Delta P_{\text{totalsuc.}}$ = Total pressure drop at suction.

Net Positive Suction Head

$$\text{NPSH}_A = P_{\text{suc}} - \text{vapour pressure}$$

$$= 2.2274 - 1.086$$

$$= 1.1414\,\text{Kg}/\text{cm}^2$$

Design margins on the discharge head are 10% of line loss +1 MLC

$$\text{Margin}_d = \frac{10^* \left(\Delta P_{\text{dis}} \right)}{100} + \frac{\text{Specific Gravity}}{10}$$

$$= 0.1161\,\text{Kg}/\text{cm}^2$$

Pump discharge head

$$P_{\text{disch}} = \Delta P_{\text{Totaldis}} + P_{\text{sdisc}} + \text{Margin}_d$$

$$= 4.161 + 1.033 + 0.1161$$

$$= 5.3101\,\text{kg}/\text{cm}^2$$

Differential pressure of the pump

$$P_{\text{diff}} = P_{\text{disc}} - P_{\text{suct}}$$

$$= 5.3101 - 2.2274$$

$$= 3.0827\,\text{kg}/\text{cm}^2$$

Calculation of differential head of the pump

$$\text{Differential head} = \frac{\text{Diff.Press.}\ ^{*}10}{\text{Sp.Gravity}}$$

$$= \frac{3.0827^{*}10}{1}$$

$$= 30.827\,\text{MLC}\,(\text{meters of liquid column})$$

Calculation of hydraulic power

$$\text{Power in kW} = \frac{\text{Diff.Press.}\ ^{*}Q}{36.71}$$

$$= \frac{3.0827^{*}35}{36.71}$$

$$= 2.939\,\text{kW}$$

$$\text{Estimated Power} = \frac{\text{Power}}{\text{Efficiency}}$$

$$= \frac{2.939}{0.6}$$

$$= \mathbf{4.898\,kW}$$

Python Code for Sample Problem 2

Program:

```python
import numpy as np
Q_n = float(input("Enter the normal flow rate (m3/h): "))
dm = float(input("Enter the fractional value of Design Margin: "))
Q = (1 + dm) * Q_n
IDsuction = float(input("Enter the value of Internal Diameter of
suction line (m): "))
IDdischarge = float(input("Enter the value of Internal Diameter of
discharge line (m): "))
rho = float(input("Enter the density of flowing fluid: "))
mu = float(input("Enter the value of viscosity of fluid: "))
eps = float(input("Enter the value of pipe roughness (mm): "))
eps = eps*10**-3
g = 9.81


def A(ID):
 A = (np.pi / 4) * (ID)**2
 return A


def V(Q, A):
 V = Q / (A * 3600)
 return V


def Re(ID, V, rho, mu):
 Re = (ID * V * rho) / mu
 return Re


def f(Re, eps, ID):
 f = 8 * ( (8 / Re)**12 + ((37530 / Re)**16 + (-2.457 * np.log((7
/ Re)**0.9 + 0.27*(eps / ID)))**16)**-1.5)**(1/12)
 return f


Asuction = A(IDsuction)
Adischarge = A(IDdischarge)
Vs = V(Q, Asuction)
Vd = V(Q, Adischarge)
Resuction = Re(IDsuction, Vs, rho, mu)
Redischarge = Re(IDdischarge, Vd, rho, mu)
fs = f(Resuction, eps, IDsuction)
fd = f(Redischarge, eps, IDdischarge)

print("Areas:", Asuction, Adischarge, "m2.")
print("Velocities:", Vs, Vd, "m/s.")
print("Re:", Resuction, Redischarge)
print("Friction Factors:", fs, fd)

print("----- Suction Line Pressure Drop Calculations -----")
```

```
nG = int(input("Enter the number of Gate Valves in Suction Line: "))
nE = int(input("Enter the number of 90 Elbows in Suction Line: "))
equiLG = float(input("Enter the value of Equivalent length per Gate
Valve (m): "))
equiLE = float(input("Enter the value of Equivalent length per 90
Elbow (m): "))
Lstraights = float(input("Enter the length of straight pipe on
Suction side (m): "))

equiLs = (equiLG * nG + equiLE * nE)
Lfits = equiLs * IDsuction
Ls = Lstraights + Lfits
delPsuctionbyg = ((fs * Ls * rho * Vs**2) / (2 * g * IDsuction))
* 10**-4
delPstrainerbyg = float(input("Enter the strainer pressure drop per
unit gravitational force acting (kg / cm2): "))
delPtotsuction = delPstrainerbyg + delPsuctionbyg
print("Total pressure drop in suction line is", delPtotsuction,
"kg/cm2.")

print("----- Discharge Line Pressure Drop Calculations -----")


nG = int(input("Enter the number of Gate Valves in Discharge
Line: "))
nE = int(input("Enter the number of 90 Elbows in Discharge Line: "))
nC = int(input("Enter the number of Check Valves in Discharge
Line: "))
nGl = int(input("Enter the number of Globe Valves in Discharge
Line: "))
equiLG = float(input("Enter the value of Equivalent length per Gate
Valve (m): "))
equiLE = float(input("Enter the value of Equivalent length per 90
Elbow (m): "))
equiLC = float(input("Enter the value of Equivalent length per
Check Valves (m): "))
equiLGl = float(input("Enter the value of Equivalent length per
Globe Valves (m): "))
Lstraightd = float(input("Enter the length of straight pipe on
Discharge Side (m): "))

equiL = (equiLG * nG + equiLE * nE + equiLC * nC + equiLGl * nGl)
Lfitd = equiL * IDdischarge
Ld = Lfitd + Lstraightd
delPdischargebyg = ((fd * Ld * rho * Vd**2) / (2 * g * IDdischarge))
* 10**-4
```

```
h = float(input("Enter the height to which discharge line is pulled
up (m): "))
delPheightbyg = h * rho * 10**-4
delpbygControlValve = float(input("Enter the pressure drop by g
along Control Valve (kg/cm2): "))
delPtotdischarge   =   delPdischargebyg   +   delPheightbyg   +
delpbygControlValve
print("Total pressure drop in discharge line is", delPtotdis-
charge, "kg/cm2.")

print("----- Suction Pressure Calculations -----")
Pssuc = 1.033 #Pressure on surface of liquid = 1 atm = 1.0332 kg/cm2
Zsl = float(input("Enter the height of liquid level from ground
(m): "))
Zpump = float(input("Enter the height of Pump Nozzle from ground
(m): "))

Psuctionbyg = Pssuc + (Zsl - Zpump)*rho*10**-4 - delPtotsuction
print("Suction pressure is", Psuctionbyg, "kg/cm2.")

VPbyg = float(input("Enter the value of Vapur Pressure by g of
Liquid (kg/cm2): "))
NPSHA = Psuctionbyg - VPbyg
DesignMargin = (0.1*delPdischargebyg) + (rho) * 10**-4

print("----- Discharge Pressure Calculations -----")
Psdisc = 1.033
Pdischargebyg = delPtotdischarge + Psdisc + DesignMargin
print("Discharge pressure is", Pdischargebyg, "kg/cm2.")

PumpPressure = Pdischargebyg - Psuctionbyg
Pumphead = (PumpPressure / rho) * 10**4

Phydraulic = (PumpPressure * Q_n) / 36.71
efficiency = float(input("Enter the value of Efficiency of pump: "))
Pestimated = Phydraulic/efficiency

print("Power (hydraulic) required =", Phydraulic, "kW.")
print("Power (estimated) required =", Pestimated, "kW.")

Enter the normal flow rate (m3/h): 35
Enter the fractional value of Design Margin: 0.1
Enter the value of Internal Diameter of suction line (m): 0.15406
Enter the value of Internal Diameter of discharge line (m): 0.10226
Enter the density of flowing fluid: 1000
```

```
Enter the value of viscosity of fluid: 0.001
Enter the value of pipe roughness (mm): 0.047
Areas: 0.018641019828626856 0.0082129931034895 m2.
Velocities: 0.5737049014894066 1.3021372731824938 m/s.
Re: 88384.97712345797 133156.55755564183
Friction Factors: 0.019904524015870586 0.019536022886823098
----- Suction Line Pressure Drop Calculations -----
Enter the number of Gate Valves in Suction Line: 3
Enter the number of 90 Elbows in Suction Line: 3
Enter the value of Equivalent length per Gate Valve (m): 13
Enter the value of Equivalent length per 90 Elbow (m): 30
Enter the length of straight pipe on Suction side (m): 6
Enter the strainer pressure drop per unit gravitational force act-
ing (kg / cm2): 0.05
Total pressure drop in suction line is 0.05560788669590639 kg/cm2.
----- Discharge Line Pressure Drop Calculations -----
Enter the number of Gate Valves in Discharge Line: 2
Enter the number of 90 Elbows in Discharge Line: 4
Enter the number of Check Valves in Discharge Line: 1
Enter the number of Globe Valves in Discharge Line: 1
Enter the value of Equivalent length per Gate Valve (m): 13
Enter the value of Equivalent length per 90 Elbow (m): 30
Enter the value of Equivalent length per Check Valves (m): 135
Enter the value of Equivalent length per Globe Valves (m): 450
Enter the length of straight pipe on Discharge Side (m): 23
Enter the height to which discharge line is pulled up (m): 20
Enter the pressure drop by g along Control Valve (kg/cm2): 2
Total pressure drop in discharge line is 4.161387845363112 kg/cm2.
----- Suction Pressure Calculations -----
Enter the height of liquid level from ground (m): 13
Enter the height of Pump Nozzle from ground (m): 0.5
Suction pressure is 2.2273921133040937 kg/cm2.
Enter the value of Vapur Pressure by g of Liquid (kg/cm2): 1.086
----- Discharge Pressure Calculations -----
Discharge pressure is 5.310526629899423 kg/cm2.
Enter the value of Efficiency of pump: 0.6
Power (hydraulic) required = 2.9395180626760147 kW.
Power (estimated) required = 4.899196771126691 kW.
```

NOTE: On changing values of various parameters in the above problem, the desired results can be obtained by using above Python code.

Appendix D: Cooling and Heating Systems (Figs. D.1and D.2, Tables D.1, D.2, and D.3)

Properties of Fuels

Solid Fuels

Solid fuels	NCV Kcal/kg	Ash %	Moisture % as per source	Sulphur content %	Volatile matter %
Coal – Anthracite	7400–7700 GCV	9.5–15. wt. %	3.0–15.0	0.6–0.8 wt. %	<10.0
Steam coal – Good grade – Lower grade	5200–5700 3400–4500	8–20 20–30	7–12 12–20	0.8–1 1.2–1.8	30–40 35–48
African coal	5500–6000 GCV	15–17	8.0–9.0	–	27
USA coal	6200–7000	10	2–15 (bituminous)	0.7–4	–
Indonesian coal	5100–6100	14–17	8.0–9.5	0.5–0.6	28–30
Indian coal	3500–4000 GCV	35–38	5–6	Upto 0.8	20–25
Petroleum coke	6800–7000	Less than 1%	10 (max)	5–7%	6–7

Notes Indian coal has much higher content of ash (35–40%) and sulphur. It is necessary to provide electrostatic precipitators and flue gas desulphurisation units for treatment of flue gases. Quality of coal from USA, South Africa, and Australia is better.

Use of petroleum coke (generally available from petroleum refineries) also needs air pollution control units due to higher sulphur content.

© The Author(s), under exclusive license to Springer Nature Switzerland AG 2022
K. R. Golwalkar, R. Kumar, *Practical Guidelines for the Chemical Industry*,
https://doi.org/10.1007/978-3-030-96581-5

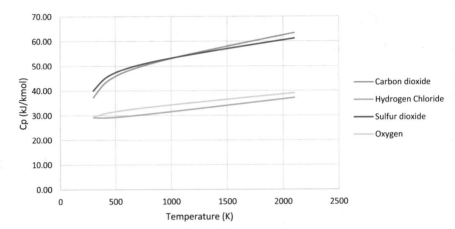

Fig. D.1 Specific heat capacity for carbon dioxide, hydrogen chloride, sulfur dioxide, oxygen

Liquid Fuels

Liquid fuels	GCV Kcal/kg	Ash %	Moisture %	Sulphur %	Boiling point range °C	Flash point °C	Pour point °C	Specific gravity	Viscosity cSt
Furnace oil	10,500 (GCV)	0.1	1	2.0–4.0	185–500	66	20	0.89–0.95	125–180 (50 °C)
Naphtha	10,600 (GCV)	–	–	0.15	97.2	−20 to −50 °C	>9	0.6–0.7	0.5–0.75 (40 °C)
Light diesel oil	10,700 (GCV)	0.02	0.25	1.0–1.8	207–535	66	18	0.85–0.87	2.5–15.7 (40 °C)
LSHS	10,600 (GCV)	0.1	1	< 0.5	185–500	93	72	0.88–0.98	50 (100 °C)
High speed diesel	10,800 (GCV)	0.01	–	0.005 max	110–375	32–96	15	0.82–0.86	2–4.5 (40 °C)
Kerosene	12,663 (GCV)	–	–	0.04	150–300	37–65	174–266	0.78–0.82	2.5 (40 °C)

Gaseous Fuels

Gaseous fuels	NCV Kcal/kg appx	Molecular weight	Sulphur %
Natural gas	11,250	16 g/mol	5.5 mg/m^3
Ethane	11,400	30.07 g/mol	–
Propane	11,000	44.10 g/mol	0.02
Liquefied petroleum gas	11,900	44.097 g/mol	<40 mg/m^3

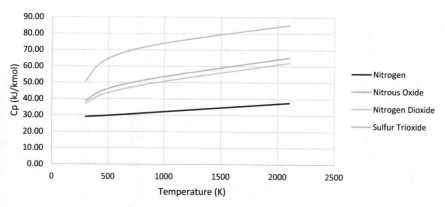

Fig. D.2 Specific heat capacity for nitrogen, nitrous oxide, nitrogen dioxide, sulfur trioxide

Table D.1 Specific heat capacity for some gases

Temp(K)	Cp (KJ/Kmol K)							
	CO_2	SO_2	O_2	HCl	N_2	N_2O	NO_2	SO_3
300	37.29	40.00	29.42	29.19	29.12	38.75	37.08	50.91
400	42.83	44.77	30.76	29.10	29.45	43.51	41.28	59.98
500	45.87	47.34	31.60	29.33	29.87	46.26	43.75	64.66
600	47.91	49.03	32.25	29.70	30.32	48.21	45.55	67.60
700	49.49	50.32	32.81	30.12	30.79	49.79	47.03	69.72
800	50.82	51.39	33.33	30.58	31.27	51.17	48.34	71.40
900	52.00	52.33	33.81	31.06	31.75	52.43	49.55	72.83
1000	53.10	53.20	34.27	31.54	32.23	53.62	50.70	74.10
1100	54.13	54.01	34.73	32.04	32.72	54.76	51.80	75.27
1200	55.13	54.79	35.17	32.54	33.21	55.87	52.89	76.38
1300	56.09	55.54	35.61	33.05	33.70	56.96	53.95	77.43
1400	57.04	56.27	36.05	33.55	34.19	58.03	54.99	78.44
1500	57.97	57.00	36.48	34.06	34.68	59.09	56.03	79.43
1600	58.89	57.71	36.91	34.58	35.17	60.14	57.06	80.40
1700	59.81	58.41	37.34	35.09	35.66	61.19	58.08	81.35
1800	60.71	59.11	37.77	35.60	36.15	62.23	59.10	82.29
1900	61.61	59.80	38.20	36.12	36.65	63.26	60.11	83.22
2000	62.51	60.49	38.62	36.63	37.14	64.29	61.13	84.15
2100	63.40	61.18	39.05	37.14	37.63	65.32	62.14	85.07

Table D.2 Properties of Commonly used Brines

Sr. No	Salt	Composition (by wt %)	Freezing Point (in °C)	Specific gravity	Specific heat capacity (J/ kg K)	Application
1	NaCl	5	−3	1.04	3900	Table salt, used in snowy regions to melt ice
		10	−6.5	1.08	3700	
		15	−11	1.119	3500	
		20	−16.5	1.16	3390	
2	KCl	5	−2.5	1.03	3900	Potash fertilizers, salt substitute for high BP patients
		10	−5	1.06	3800	
		15	−7.5	1.099	3550	
		20	−10	1.14	3200	
3	CaCl$_2$	10	−6.7	1.045	3900	Waste water treatment, drying air and gases
		20	−20	1.085	3600	
		30	−47	1.125	3300	
		40	16	1.19	3000	

Table D.3 Properties of commonly used Refrigerants

| Name of refrigerant | Vapor pressure (kPa) at various temperatures (°C) | | | | | | Toxicity | Corrosive nature | Boiling Point | Latent heat of vaporisation (KJ/kg) |
	-20	-10	0	10	20	30				
Ammonia (R 717)	190.08	290.71	429.38	615.05	857.48	1167.2	Higher concentration can have adverse effects on eyes, skin and lungs	Liquid ammonia containing impurities corrodes mild steel	−33.34 °C	1369
Chlorodifluoro methane (R 22)	245.31	354.79	497.99	680.95	910.02	1191.9	Burns similar to frostbite.	It releases corrosive gases when subjected to adverse conditions.	−40.8 °C	233.95
Pentafluoro ethane (R 125)	337.33	482.52	670.52	908.75	1205.2	1568.5	At high concentrations, inhalation causes headaches and drowsiness.	No significant damage to equipment	−48.5 °C	165
Ethane (R 170)	1421.5	1858.8	2386.7	3017.2	3765.5	4655.1	Exposure may cause headache, nausea, vomiting, dizziness, and lightheadedness. Very high levels cause suffocation	Prolonged exposure to fire and heat causes rupture of containers	−88.5 °C	487.64
n-butane (R 600)	45.21	69.55	103.23	148.45	207.65	283.41	Inhalation can cause euphoria, drowsiness, unconsciousness, asphyxia, cardiac arrhythmia, fluctuations in blood pressure, and temporary memory loss.	No corrosive action towards metals	−1 ° C	386

Conclusion

Chemical industries are very important for making available products required for daily needs and for smooth comfortable life; for health care of the people and for economic progress of the country. It is necessary that these industries are run in a safe and efficient manner without causing any environmental pollution.

This book has been written with an aim to provide guidelines for selection of process, identifying hazards, ensuring preventive steps, well-designed equipment (with proper materials of construction) and plant layout, and engaging qualified, experienced manpower for carrying out the operations to run the organisation in a sustainable manner.

Check lists are provided for project and plant managers for the revival of idle plants, relocation and modernization of plants, and for carrying out the more common operations like pumping of fluids, heating and cooling systems process control, reducing costs of production, and environmental pollution control.

It is hoped that these will be found useful by readers associated with chemical industries.

© The Author(s), under exclusive license to Springer Nature Switzerland AG 2022   287
K. R. Golwalkar, R. Kumar, *Practical Guidelines for the Chemical Industry*,
https://doi.org/10.1007/978-3-030-96581-5

Index

A
Actions to prevent breakdowns
 before and during commissioning
   runs, 237–238
 check operation, 238–239
 at design stage, 233–234
 during erection, 235–237
 during procurement, 234–235
 during testing of process units and
   mechanical trials of machinery, 237
Additional precautions, 79
Advantages of process control, 168
Agitated steam-jacketed reactor, 21
Air compressor, 172
Air pollution control equipment, 146
Alloy 20, 36
Alloy steel, 34
Anodic protection, 46
Arranging funds for various activities, 2, 3
Arranging site infrastructure, 7
ASME/ANSI codes
 piping, 264
 pressure vessels, 259–261
ASME Boiler, 59
Austenitic steels, 36
Axial piston pumps, 108

B
Ball valves, 89, 90
Basic engineering package (BEP), 4
Boiler and pressure vessel code (BPVC)
 corrosion, 63
 design pressure, 63
 divisions, 61
 fired or unfired, 62
 rules, 60, 61
 welding variables, 62
Boiler feed water (BFW), 150
Breaking pin device, 78
 explosion hatches, 79
 liquid seals, 79
Butterfly valves, 89

C
Calculated allowance, 193
Capacity expansion of existing plant, 188
 augment existing process units and
   machinery, 189–190
 each sub-system, checking, 188
 practical guidelines, 188
 revival of an old/idle plant, 190
   decontamination, 190
   general, 191–192
   servicing and replacement/repairs, 191
Casing, 115
Cathodic protection (CP), 44, 45
Cavitation, 119
Centrifugal pumps, 112
 advantages, 116
 basic principle, 114
 Bernoulli's principle, 114
 capacity expansion, 118, 119
 cavitation, 119
 disadvantages, 117
 energy conversion, 114
 hot liquids, 120
 impeller selection, 117
 installation, 123, 126, 127